WHAT THE EYES
DON'T SEE

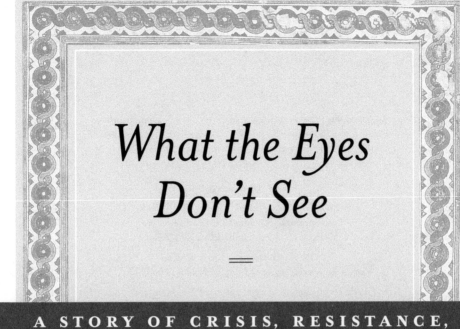

What the Eyes Don't See

═

A STORY OF CRISIS, RESISTANCE,

AND HOPE IN AN AMERICAN CITY

═

MONA HANNA-ATTISHA

ONE WORLD | NEW YORK

Copyright © 2018 by Mona Hanna-Attisha, M.D.

Published in the United States by One World, an imprint of Random House, a division of Penguin Random House LLC, New York.

ONE WORLD is a registered trademark and its colophon is a trademark of Penguin Random House LLC.

Grateful acknowledgment is made to Liveright Publishing Corporation for permission to reprint nine lines from "A Worker's Speech to a Doctor," translated by Thomas Mark Kuhn (originally published in German in 1939 as "Rede eines Arbeiters an einen Arzt"), from *Collected Poems of Bertolt Brecht* by Bertolt Brecht, translated by Thomas Mark Kuhn and David J. Constantine, copyright 1939, 1961, 1976 by Bertolt-Brecht-Erben/Suhrkamp Verlag, copyright © 2016 by Thomas Mark Kuhn and David J. Constantine. Used by permission of Liveright Publishing Corporation.

Image credits appear on page 365.

Library of Congress Cataloging-in-Publication Data
Names: Hanna-Attisha, Mona, author.
Title: What the eyes don't see : a story of crisis, resistance, and hope in an American city / Mona Hanna-Attisha.
Description: New York : One World, 2018. | Includes bibliographical references.
Identifiers: LCCN 2018002721| ISBN 9780399590832 (hardback) | ISBN 9780399590849 (ebook)
Subjects: LCSH: Lead poisoning—Michigan—Flint. | Drinking water—Lead content—Michigan—Flint. | Water quality management—Michigan—Flint. | Hanna-Attisha, Mona. | Physicians—Michigan—Flint—Biography. | Flint (Mich.)—Environmental conditions. | BISAC: SCIENCE / Environmental Science. | MEDICAL / Public Health. | SOCIAL SCIENCE / Sociology / Urban.
Classification: LCC RA1231.L4 H34 2018 | DDC 615.9/256880977437—dc23 LC record available at https://lccn.loc.gov/2018002721

Printed in the United States of America on acid-free paper

randomhousebooks.com

246897531

First Edition

Book design by Barbara M. Bachman

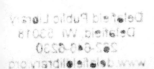

For the children
of Flint

Unless someone like you
cares a whole awful lot,
nothing is going to get better.
It's not.

—DR. SEUSS, *The Lorax*

Contents

===

WHAT THE EYES
DON'T SEE

How I Got My Name

I AM IRAQI, AN IMMIGRANT, BORN SOMEWHERE ELSE—BUT NOT in Baghdad like my older brother, who was named Muaked, which means "certain, confident." He fits that name, always did. He was just one year old when my family moved to Sheffield, England, and Muaked's name proved difficult for English speakers to pronounce—just as they had trouble with my dad's name, Muafak, which sounds like a profanity even if you say it correctly.

In England, my family stayed for a time with my mom's cousin Bertha, who was born in Iraq but over the years had become British to her core. In a display of her extraordinary strength of personality, she renamed my brother "Mark," and it stuck.

MY GRANDFATHER HAJI CAME up with my name. Haji was idolized for his charm, intelligence, and humanity. He was a businessman who lived in Baghdad and had a large, soulful view of the world, an iconoclastic wisdom. People in our family always wanted him to name their babies, probably for that reason.

He named me Mona because he thought it would be an easy name for both English and Arabic speakers to pronounce. My fam-

ily was still living in England at the time but was planning to return
to Iraq when my dad's studies at the University of Sheffield ended.
He was getting a doctorate in metallurgy, which is what you study
if you are going to work on nuclear power plants—or nuclear weapons. But my dad was a progressive, a pacifist, and didn't want his
work to go toward making weapons for the repressive Ba'athist regime that dominated Iraq. He was interested in working with metals like zinc and aluminum and in creating new alloys. He has
an engineer's passion for making things work better—sometimes
stronger, sometimes lighter, sometimes more durable.

In Arabic, my name is traditionally spelled and pronounced
Muna. But Haji believed that, for me, the anglophone version, with
a long *o*, was better. Haji was magical enough that maybe he fore-
saw that a Western name would work to my advantage. Either way,
Mona means "hope, wish, or desire."

I was a chubby baby, born with a mark, a capillary hemangioma,
on my forehead. It wasn't pretty or fascinating, like Harry Potter's
lightning bolt. My mark was dark red, the size of a golf ball, and
near my hairline. Sometimes it would bleed when I fell. When my
mom, Talia, carried me in public, the women of Sheffield looked
at me with horror and pity and sometimes got up their nerve to

ask: What's that growing on your baby's forehead? *Can it be removed?*

The hemangioma regressed—went away on its own—as I grew older. Now it is just a spot where my hair never grows. Your eyes wouldn't see it unless I told you where it was.

My brother's real name also vanished over time. My mom is the only one who calls him Muaked anymore. I don't think his own kids even know it.

The road behind my family disappeared too. The Iraq they knew was lost, replaced by war and ruins. In my mind, this lost Iraq is a land of enchantment and despair. But its lessons endure. They may be unseen, but they are not forgotten, just as Mark is still Muaked—and will always be certain and confident. And when I touch my forehead, my birthmark is still there.

THE HIGHWAY WAS DARK, and snow was falling but not much. My mom drove for a few hours and then stopped and switched places with my dad. It was an endless drive, close to twelve hours, but we'd made the trip once before and had our routines. We were going back up north, to Houghton, after spending Christmas with Mama Evelyn, my paternal grandmother, in Southfield, just outside Detroit.

My dad put the silver Monte Carlo into cruise control at sixty-five. My mom resumed her knitting and turned on the radio. Mark and I were sitting in the deep sofalike backseat. Nobody was wearing seat belts. We'd long ago tucked them underneath the seats to keep them out of the way.

We passed Flint, where the Monte Carlo had probably been built. We passed Saginaw and the landscape soon became wilder and more pristine. My brother played with the Legos he had gotten for Christmas. I had a soft stuffed bear on my lap, a present from Santa, the only one I'd wanted.

My dad opened the window to flick the ash of his cigarette into the frozen air. When he rolled it back up, I could hear the clacking of my mom's knitting needles again. On the way down to Mama Evelyn's, she'd made a cozy oversize sweater for me. Now on the way back, she was finishing one for Mark.

My mother was perpetually busy—frying *kibbee* on the stove, pulling baklava trays out of the oven, rolling a new batch of dolma. She sewed most of our clothes. At night, she was always reading a book. In Iraq, she had been trained as a chemist—one of only two women in her chemistry classes at Baghdad University—but her foreign degree was all but meaningless in America. Never mind that, though. We would go back to Iraq someday soon, she always said, her voice thick with emotion and defiance. Back to Baghdad.

She was raised in the multicultural Al-Jadriya Karrada district, along a gentle bend of the Tigris River, and never stopped dreaming of the river's *masgoof,* the carp that were butterflied, then grilled upright in riverside parks—the Iraqi national dish. She dreamed about the flavors and spices of home, the gardens of citrus and date trees and sweet-smelling honeysuckle and gardenia, the soft, mesmerizing music and dry heat. She missed her parents, her brothers and sisters, and the fragile beauty of her Baghdad.

At bedtime, she regaled us with stories of the ancient capital, once the most advanced, prosperous, and progressive civilization in the world—the center of mathematics, astronomy, and medicine. Baghdad was where algebra was invented, where *The Canon of Medicine* was written, a medical textbook that was taught throughout the known world for six centuries. She would weave her recounting of Mesopotamia's history with strands of mysticism and fable, the romantic tales of Sinbad, Ali Baba, and Aladdin as told by Shahrazad, which was also her youngest sister's name.

In that pantheon of magicians and heroes, she included Haji—her father, who gave me my name. Haji sent presents at Christmas. He called to speak to us on the phone in his throaty voice, deepened by years of cigar smoking and scotch drinking. He spoke colloquial Baghdadi—and his words and intonations and accent all felt like a secret language, as if we were the only family on earth who understood it.

Someday soon, my mom said, we'd see Haji again, and our grandmother Mama Latifa, an elementary school teacher and, according to family lore, the best cook who ever lived. Her voice on the phone was the sound of deep love. She wanted to hug us, to cuddle us, and couldn't wait to have us with her. But my dad said it wasn't safe to go back, even for a visit.

As soon as I was old enough to understand the conversation passing between my parents, I heard things that were hard to process, things that haunted my nightmares. It was in these early years that I first discovered the concept of evil—not from cartoon monsters or Hollywood villains but from hearing about the right-wing fascist who had risen to power in Baghdad, a political ascension by way of thuggery and corruption, torture and murder.

Every night my father listened to his Grundig shortwave radio, which crackled with static and delivered news of Saddam Hussein's brutality and violence. My dad was always fine-tuning the knobs of the radio, trying to pick up a faraway station, desperate to hear something through the static. Friends and relatives, fellow dissidents, told him about people who had disappeared from their homes and later were released from prison with broken bones, crooked limbs, and burn marks. Sometimes a mutilated body was found on a doorstep. Sometimes the disappeared never came back at all.

Something new, something terrible, was always keeping us from returning to Iraq. Saddam's reign became more extreme every year. My parents knew their letters home were opened by the Mukhabarat, the secret police, and we assumed that all phone calls to my mom's parents were monitored. Iraq was called a "Republic of Fear," a place where neighbors feared one another and parents feared even their own children: a haphazard comment made to a schoolteacher could lead to torture and prison. So we were waiting it out. Waiting for Saddam's evil cult of personality to be brought down. Waiting for our home to be restored to a just and peaceful country, a republic in more than name. We weren't the only ones; Mama Evelyn and three of my dad's siblings moved to Metro Detroit along with thousands of displaced Iraqi immigrants who were settling there. It was only the beginning of the Iraqi diaspora.

From six thousand miles away, my dad grew more vigilant and outspoken. Injustice gnawed at him, changed him. He was gruff, hard to hug, and undemonstrative with us. He holed up at his desk for hours, listening to the shortwave, smoking cigarette after cigarette, channeling his anger into opposition. Sometimes when my mom put us to bed at night, the loneliest look came over her face and she turned away, not wanting us to see her cry.

THAT CHRISTMAS, MAMA EVELYN had taught me how to play Konkan, an Iraqi variant of gin rummy that covertly doubles as a way to teach kids arithmetic. I picked up the game quickly. Mama Evelyn was pleased about that. She liked having more Konkan players and having quick and curious kids in the family. Mark and I sat for hours on the floor of her tiny apartment, shuffling and dealing out fourteen cards to start, then were sometimes joined by my uncles Muthefur and Munathel or the stray relative who was, like us, in exile from Iraq. Our hands were still too small to hold all fourteen cards.

Nobody had money for a hotel or for much in the way of Christmas presents. We camped together at Mama Evelyn's, sleeping in the living room, piling our suitcases and clothes and toys in the corner, next to the grocery bags full of Arabic grains and spices that my mom had bought in Detroit to take back up north.

Mama Evelyn was a tough competitor, which always made Konkan more compelling. She was scary-smart and strong-willed—a force of nature—and for a grandmother, pretty young. She'd married at fifteen, or around then, no one really knows. She had my father, her oldest child, not long after.

Her late husband, my dad's dad, Dawood Hanna, was a railroad station manager who moved his family ten times between 1940 and 1963. He followed his government job, always living in houses next to the railroad stations. When they lived in Basra—the historic port city of southern Iraq, along the Shatt-al-Arab, where the Tigris and Euphrates rivers merge—my dad had a pet monkey named Maimun. I loved hearing about Maimun and begged Mama Evelyn for the details. And next door to the Basra train station, there was a house filled with exotic birds, an aviary of some kind. They were brightly colored and had strange squawks and screams. I asked Mama Evelyn why anybody would have a house of birds. What was it for? It was a mystery, she said, a nonsensical moment of beauty, a colonial creation from the days when the Kingdom of Iraq was a puppet state run by the British.

Mama Evelyn was confident, never too troubled by small or large setbacks, and always positive, even when her beloved Detroit Pistons lost. She never talked at all about her biggest life tragedy—when my grandfather was fired from the railroad job he loved and sent to prison by the Ba'athists during a political purge of leftist civil servants. He came home again but was never the same—and he died just a few months before I was born. Mama Evelyn wore only black, in mourning, for the next ten years—including that Christmas.

MY DAD WAS SILENT as he drove. The road became darker and narrower as the night descended. There wasn't much traffic. The radio signal grew fainter, and my mom switched over to the eight-track player and we listened together to the soulful voice of Fairuz. My mom began singing along with the legendary Lebanese singer, their delicate voices blending into a duet.

I was hoping to stay awake until we crossed the Mackinac Bridge, the longest suspension bridge in the Western Hemisphere. But I began to drift off, feeling cozy and secure inside the Monte Carlo, our lumbering tank, heavy, with a smooth ride. It was marketed as high-end and named to make people think of Monaco and Princess Grace, but it was basically a tank with a big V8 engine, built to go fast and take some damage.

My dad bought the used 1975 Monte Carlo himself. We weren't a General Motors family yet—it wouldn't be until later that we got a free loaner car every three months that my dad was technically supposed to test-drive and "rate." That year he was still a postdoctoral researcher at Michigan Tech University, continuing his metallurgical research in Houghton, in the Upper Peninsula, or "Copper Country," and struggling to raise a family on a small stipend.

Our first winter, almost three hundred inches of snow fell in Houghton. It's a testament to human adaptability that my parents acclimated from the dry heat and palm trees of the Middle East to the dreary days of England to the frozen winters of Michigan's Upper Peninsula. The north wind chilled our tiny rented house on Portage Lake, and the Monte Carlo sported a permanent layer of snow on the roof. But we were warmed by the kindness of Upper Peninsula Michiganders—who cheerfully called themselves "Yoopers"—and the tight-knit international graduate student community.

I started elementary school in Michigan when I was four, after my parents mistakenly wrote my birth date in European style on the school forms; it was misunderstood as September 12 rather than

December 9, which meant I was always the youngest in my class. Just before Christmas at Mama Evelyn's, I turned five. Thanks to the cartoons I watched on TV and my kindergarten class at Houghton Elementary, when my parents asked me a question in Arabic, I did a new thing.

I answered in English.

WE WERE DRIVING THROUGH Gaylord, just fifty miles south of the Mackinac Bridge, when our car hit a patch of black ice. Before I could react or even call out, the Monte Carlo began swerving and sliding. It slammed into the metal guardrail and spun away, crossing the lines in the middle of the road and veering onto the opposite lanes of the highway, where it smashed into another guardrail and kept spinning.

I was flying, as if I were weightless, from one side of the car to the other. Our car kept grinding against the guardrail until the guardrail ran out and there was nothing left to break our slide. The unstoppable Monte Carlo fell headlong into a ravine, crashing and crushing trees on its way down.

I have a faint memory of flashlights and an ambulance. My mom was crying and pulling me into the front seat, onto her lap. I saw blood on my dad's forehead. I blacked out.

The next morning I woke up in the pediatric wing of a hospital in Traverse City. I couldn't move my neck. I couldn't really move at all. The pain was terrible, but I remained silent, afraid to make a noise. My mom was trying to communicate with the hospital staff in her limited English, which did not include any medical terms. She was worried about my mouth. Something didn't look right to her. My face seemed crooked.

A young woman in a white coat quietly entered the room. She listened closely to my mom. Then she came to the side of my bed and looked at me. She had brown skin and dark hair, like me. She smiled a big smile. She held my small hand firmly in hers. She told me I was going to be okay.

It took a while for me to get better. A spinal injury. A broken jaw. I had some operations. My neck was in some sort of brace. My top and bottom teeth were wired together. My mom slept with me at the hospital. Once back home, she put rice and stews, *timen wa maraca,* in a blender and turned them into smoothies that I could drink through a straw. I missed a month of kindergarten. The sturdy Monte Carlo was totaled. But I was okay.

SOME THIRTY-PLUS YEARS LATER, I'm wearing the white coat. I'm smiling at the beautiful brown girl in front of me and firmly holding her hand. She's going to be fine, I tell her. She was mixed up in an accident. A lot of kids were. The accident wasn't her fault. And it is my job to make sure she is okay.

THIS IS THE STORY of the most important and emblematic environmental and public health disaster of this young century. More bluntly, it is the story of a government poisoning its own citizens, and then lying about it. It is a story about what happens when the very people responsible for keeping us safe care more about money and power than they care about us, or our children.

The crisis manifested itself in water—and in the bodies of the most vulnerable among us, children who drank that water and ate meals cooked with that water, and babies who guzzled bottles of formula mixed with that water. The government tried hard to convince parents the water was fine—safe—when it wasn't. But this is also a story about the deeper crises we're facing right now in our country: a breakdown in democracy; the disintegration of critical infrastructure due to inequality and austerity; environmental injustice that disproportionately affects the poor and black; the abandonment of civic responsibility and our deep obligations as human beings to care and provide for one another. Along with all that—which is a lot already—it's about a bizarre disavowal of honesty, transparency, good government, and respect for scientific truth.

Those are demoralizing realities to face. But there is another story, another side of Flint. Because it is also a story about how we came together and fought back, and how each of us, no matter who we are—a parent, an activist, a schoolteacher, a pediatrician—has within us a piece of the answer. We each have the power to fix things. We can open one another's eyes to problems. We can work together to create a better, safer world, a place where all children can develop without obstacles and barriers, without poisoned water or callousness toward their dreams.

There are lots of villains in this story. A disaster of this scale does not happen completely by accident. Many people stopped caring about Flint and Flint's kids. Many people looked the other way. People in power made tragic and terrible choices—then collectively and ineptly tried to cover up their mistakes. While charges have been brought against some of the individuals who were culpable, the real villains are harder to see.

Because the real villains live underneath the behavior, and drive it. The real villains are the ongoing effects of racism, inequality, greed, anti-intellectualism, and even laissez-faire neoliberal capitalism. These are powerful forces most of us don't notice, and don't want to. These villains poisoned Flint with policy—with decisions that were driven by lack of hope in government. If we stop believing that government can protect our public welfare and keep all children safe, not just the privileged ones, what do we have left? Who are we as a people, a society, a country, and a civilization?

For all the villains in this story, there are also everyday heroes: the people of Flint. Each one has a story to tell—100,000 stories in all—about months of pain, anger, betrayal, and trauma, along with incredible perseverance and bravery. Flint fought hard, never gave up, and turned a devastating crisis into a model of resilience. But this book is only my story, told from my narrow perspective, as a doctor and as a brown immigrant in a majority-black city. It cannot attempt to do justice to all the stories that need to be told. No one book could.

But I will share with you a few stories of my Flint kids. They are my inspiration. To protect their privacy and dignity, I have changed or modified their names and identities. In some cases, composites have been created that are based on real patients and real encounters. They are strong, smart, beautiful, and brave—and so resilient.

Resilience isn't something you are born with. It isn't a trait that you have or don't have. It's learned. This means that for every child raised in a toxic environment or an unraveling community—both of which take a terrible toll on childhood development and can have lasting effects—there is hope. This is another way we can come together and each be a piece of the answer, not just for Flint and places like Flint, but for children anywhere who bear the brunt of life's hardest blows, and live with poverty, violence, and hopelessness. Resilience is the key, the deciding factor between a child who overcomes adversity and thrives and a child who never makes it to a healthy adulthood.

Just as a child can learn to be resilient, so can a family, a neighborhood, a community, a city. And so can a country. A country can endure trauma and neglect and become a place where people are cared for, where democracy and equality and opportunity are once again encouraged and advanced. Where poverty is silenced instead of people. Where we nurture one another and create stable and safe environments for all children to grow up.

This is where healing begins.

CHAPTER 1

What the Eyes
Don't See

I T WAS COLD AND RAINY ON THE SUMMER MORNING OF AUGUST
26, 2015—that predictably unpredictable Michigan weather. I
dropped my two daughters off at Skull Island, an ominously named
summer camp they were trying out for a few days, and got to Hurley later than usual. Once inside, warm, musty air greeted me, that
old-hospital smell wafting through the brightly lit corridors of
glossy tile.

I reached my office on the pediatrics floor and immediately felt behind the door for my white coat. Wearing the coat always made me feel better, stronger, protected—it was my armor. Then I swung a stethoscope around my neck, completing my transformation from civilian to doctor.

At my desk, I read the local news online for a few minutes—just a quick scan—and noticed another story about the tap water in Flint. Residents were complaining, authorities were explaining. This had been going on for so long, more than a year already. It had become a loop of white noise. Turning my attention to the multi-colored grid of my online calendar, I got a handle on the day ahead. It was packed—a crush of meetings, four, five, six of them, for research projects, curriculum changes, faculty recruitment. Somehow, wedged between the meetings, I had to answer emails and read other material, mostly prep for more meetings.

On the home front, a barbecue was beginning to materialize for later that night—a spontaneous, last-minute gathering. I'd discovered that my high school friend Annie Ricci was in town for a few days, and we'd decided to get together with another old friend, Elin Warn (now Betanzo). We had all been friends since our freshman year at Kimball High School in Royal Oak, the inner-ring working-class Detroit suburb where my family had landed after my dad was hired by General Motors. Annie is an opera singer in New York City now. Elin is an environmental engineer who moved back to the Detroit area after many years in D.C. They were both in my wedding, but not as bridesmaids in matching dresses. I didn't have any of those. Elin and Annie were both serious musicians, so I asked them to perform—and they did.

I reached into the pocket of my white coat for a pen and felt the plastic tip of an otoscope, that pointy tool that doctors use to look inside a child's ear. The coat may have been my soft cotton armor, but as far as I can tell, the real reason doctors wear white coats is for the pockets. Otherwise there'd be no place to store all our pens, pagers, cellphones, reference cards, tongue depressors, penlights, mints, Chapstick, and otoscope tips.

My fingers felt a scrap of paper, and I pulled it out. The paper was covered in crayon scribbles. A memento. I remembered where it had come from and smiled. Reeva had given it to me. She was a watchful two-year-old who had been coming to our clinic since she was born at Hurley Medical Center.

The week before, Reeva's four-month-old baby sister, Nakala, had been in the Hurley Children's Clinic for a routine checkup. One of my pediatric residents, Allison Schnepp, was seeing Nakala; I was the supervising physician. Nakala's mom, Grace, a young African-American woman with a steady gaze and hair pulled under a loose cap, told us she wanted to stop breastfeeding. I urged her to continue, but she said she'd made up her mind. Breastfeeding took too long and was a hassle—plus, she had to go back to work. She was a waitress, and there was no place to pump in the restaurant except the restroom that all the customers used. She couldn't afford to do anything that jeopardized her job. As it was, she wasn't getting enough hours to make ends meet.

She planned to switch to powdered formula mixed with water but had some concerns. "Is the water all right?" she asked, looking skeptical. *"I heard things."*

Reeva walked toward me with her hand out. Kids love to distract a doctor who is giving total focus to a younger sibling. So I turned my full attention to Reeva, and she placed the torn scrap of paper in my open hand. She had a sheepish smile, as if she were handing me a secret message, and we shared a conspiratorial look.

"Thank you, Reeva," I whispered. "I'm going to keep this right here in my pocket." Then I sat her down on my lap.

The water. I'd been asked about it before.

"Don't waste your money on bottled water," I said, nodding at Grace with calm reassurance, the way doctors are taught. "They say it is fine to drink."

Inches away, Reeva watched me carefully. I smiled at her again, gave her an extra squeeze, and then put her down.

I patted Nakala's fuzzy little head and touched her fontanel, or soft spot, out of habit, to check its size. It was my chance to explain to Grace that her baby's skull was open because her brain was still growing. This was the time to stimulate her baby by singing, talking, and reading. Then I gave Grace another nod.

"The tap water is just fine."

I WANTED TO BE a doctor as far back as I can remember, maybe from obsessively watching *M*A*S*H* reruns growing up. Or it could've been the story about my grandfather Haji that my mom used to tell me, when he fell out of a tree and doctors took care of his broken leg. Or maybe it was the car accident and my early experience with a caring physician who made it seem like everything was going to be okay. My parents are both scientists who raised us to love the multiplication and periodic tables and the majestic order of the natural sciences, so the prospect of biology, chemistry, and math courses never put me off.

In high school, I had some powerful experiences as an environmental activist, so I created an environmental health major at University of Michigan's School of Natural Resources and Environment, merging environmental science and premed courses. My passion for activism, service, and research was solidified there, followed by four years of medical school at Michigan State University; my last two clinical years were in Flint.

It wasn't a tough call which specialty I'd go into. As a medical student, you have to do rotations in a variety of fields, and as soon as I got to pediatrics, I felt like I was at home. Kids are usually looking for fun, and everything's new to them. No matter how sick they are, they still want to laugh and play. I was briefly tempted by obstetrics but noticed that as soon as my patients gave birth, I tended to forget about them: my attention suddenly shifted to their newborn babies.

I may not be quite as much of a baby-whisperer as my husband, Elliott, who's a pediatrician like me but with the supernatural power to soothe any child. I can hold my own, though. With patience and empathy, and sometimes with stickers, bubbles, and penlight tricks, I can get that sulky five-year-old to tell me where it hurts. A non-verbal teenager will talk to me if I take her seriously and listen carefully and let her know that I am *her* doctor and not her parents'. Even when a baby is wailing and making those supersonic ear-piercing sounds, I know that it just takes a soft voice, a gentle sway, and eye contact to calm them.

A crying baby gives me a sense of mission. Deep inside I have a powerful, almost primal drive to make them feel better, to help them thrive. Most pediatricians do. For some of us, that sense of protectiveness becomes much more powerful when the baby in our care is born into a world that's stacked against her and her needs aren't being met—a world where she can't get a nutritious meal, play outside, or go to a well-functioning school, all of which will diminish her health. I am a fanatic when it comes to protecting all kids, but when I see a child in danger through no fault of her own, I go a little mad. She's a baby like any other, with wide eyes and a growing brain and vast, bottomless innocence, too innocent to understand the injustice of her circumstances. She can't see what I can.

A baby who is properly fed and loved and kept healthy, and surrounded by people and communities that value and protect her, has the best chance of becoming a healthy adult. This is what drew me to pediatrics—we pediatricians are at the pivotal intersection of clinical care and prevention. Every aspect of my job—from immunizations to emphasizing the importance of bike helmets—is not just about ensuring kids are healthy today. It's about tomorrow, next year, and twenty years from now. We see life at its beginning, when it can be shaped for good. As Frederick Douglass said, "It's easier to build strong children than to repair broken men." Walk into the adult floor of a hospital any day, and you'll see beds of patients with problems like diabetes or heart disease that can't be fixed, because to do that you'd have to time-travel back to their childhoods and fix those too.

As MUCH AS I love spending time with kids and seeing one little patient at a time, I wanted to have as much impact as I could—on as many lives as I could—so right from the beginning, I made a tactical decision to be a medical educator rather than a pediatrician in private practice. That way, over the course of my career, I could share my passion for children's health and proven interventions with hundreds of new doctors who would go on to treat thousands of young patients, caring for them as I would and hopefully even better.

And in 2011 I became the director of the pediatric residency program at Hurley Medical Center, a public teaching hospital affiliated with Michigan State University; with more than four hundred beds and almost three thousand employees, it's a place where new doctors are trained and most of the children in Flint are treated.

The hospital was given to the city of Flint by a soap and sawmill businessman, James J. Hurley, in 1905. And like many public hospitals, it serves a poor and minority population with high levels of Medicare/Medicaid patients and uncompensated charity care. That means budget cuts from the state or federal government hit Hurley hard, in ways they would not hit a private hospital.

When I first took the job in 2011, the pediatric residency program was in tough shape and coming up short in lots of ways, large and small. Morale was low. The sixty-year-old program was at risk of losing accreditation, which meant it could close altogether. Its clinic, where Flint kids came for routine appointments, was in a depressing old building with low ceilings and little sunlight.

The first things I did were to increase the number of our residents and faculty and to overhaul our programs and recruitment practices to attract better trainees. We worked hard to improve residents' curricula and schedules—and soon we received a full ten-year accreditation. When the lease was up on our old clinic location, we moved our pediatric center into a one-of-a-kind building with soaring ceilings and spectacular sunlight, built above a year-round farmers' market—and just a few steps from the central bus stop. It was a dream location: the light, the fresh produce, the beauty of the building itself. It was a chance to give the kids of Flint a glimpse at what a healthy environment might look like—but also to show them that they deserved nothing less.

On that August day, we had been in the new Hurley Children's Clinic for only a few weeks and were still settling in. But I was looking ahead already, to September 15, when applications would come in from next year's residents.

Recruitment can be difficult when your program is in Flint. Top medical residents want to live in Chicago, San Francisco, Boston, or New York. Luring them to an economically troubled community like Flint takes powers of persuasion, finesse, and assurances that they will be bountifully rewarded, but in ways that are as spiritual and personal as they are practical. But it works. Each year the residents we attract are better, more competitive, and more committed.

So while it's true that as a residency director, I don't get to care directly for kids as much as I want, I get to spend most of my days with a group of smart, compassionate young doctors who love kids—and believe in Flint—as much as I do.

PEDIATRIC RESIDENCY TAKES THREE YEARS. Each of those years is divided into four-week rotations called "blocks," and each block is focused on a different pediatric skill. I direct a rotation called Community Pediatrics, designed specifically to open the eyes of first-year residents to the powerful, but not always immediately apparent, environmental and community factors that affect the lives and health of their patients. There's an expression I have always liked, a D. H. Lawrence distillation: *The eyes don't see what the mind doesn't know.* The first time I heard that phrase was during my own pediatric residency, when it was uttered by Dr. Ashok Sarnaik, a legendary pediatrician at Children's Hospital of Michigan in Detroit. He challenged residents to know every possible disorder or genetic syndrome under the sun and its underlying pathophysiology. When discussing a case and trying to figure out a diagnosis, he watched us run through our limited supply of options, and he always criticized us for not reading enough and therefore not knowing enough, for not seeing the whole picture.

"How can your eyes see something," he'd say, "that your mind doesn't know?"

Community Pediatrics is meant to widen the focus of pediatricians beyond whatever is immediately visible. Sure, a nosebleed is a nosebleed. An ear infection is an ear infection. But beyond the common fevers and colds, many children are facing other struggles.

Compared to nationwide averages, Flint families are on the wrong side of every disparity: in life expectancy, infant mortality, asthma, you name it. Flint is a struggling deindustrialized urban center that has seen decades of crisis—disinvestment, unemployment, racism, illiteracy, depopulation, violence, and crumbling schools. Navy SEALs and other special ops medics train in Flint because the city is the country's best analogue to a remote, war-torn corner of the world.

The city compares badly not just to the rest of the country but to neighboring communities. The median household income is half the Michigan average, and the poverty rate is nearly double. The more adversities a child experiences, the more likely she will grow up to be unhealthy in ways that are completely predictable.

A kid born in Flint will live fifteen years less than a kid born in a neighboring suburb. *Fifteen years less.* Imagine what fifteen years of life means. In a country riven by inequalities, Flint might be the place where the divide is most striking.

This is why the routine work of pediatrics—immunizations and well-baby care and the rest—is not enough for a child in Flint. Our children need much more than routine primary care just to get an even shot at the rest of their lives.

I give my Community Pediatrics residents this Bertolt Brecht poem from 1938, "A Worker's Speech to a Doctor," which lays out the stakes better than I ever could:

When we're sick, we hear
You are the one who will heal us.
When we come to you
Our rags are torn off
And you tap around our naked bodies.
As to the cause of our sickness
A glance at our rags would
Tell you more. It is the same cause that wears out
Our bodies and our clothes.

Physicians need to be trained to see symptoms of the larger structural problems that will bedevil a child's health and well-being more than a simple cold ever could. But these problems are harder for even a well-trained physician to identify. A child doesn't come to my exam room for "food insecurity." Their moms don't call the clinic for an appointment because "we can't make ends meet" or "there aren't any safe places to play outside." They make appointments because of nosebleeds and ear infections, like other moms, or for well-baby checkups. And when we see them, if we don't ask about the situation at home or learn to notice the clues on our own, we'll never find out what these larger problems are. When we know about the child's *environment,* we can treat these kids in the best, most holistic way, which will leave them with much more than just a prescription for amoxicillin.

Years ago we talked about these environmental factors as "social determinants of health." Today we call them "adverse childhood experiences" (ACEs) or "toxic stresses." These new concepts take things a step further than the old model in two ways. First, they emphasize the importance of adversity in the developmentally vulnerable window of early childhood. A child's first years are the most critical in her development and set her up for the rest of her life. It's crucial to understand this. The other new concept is our realization that a child's neuro-endocrine-genetic physiology *can be altered.* Prolonged, extreme, and repetitive stress or trauma—due to exposure to an ACE, including poverty, racism, violence—chronically activates stress hormones and reduces neural connections in the brain, just at the time in a child's development when she should be growing new ones.

In a landmark study analyzing the health data of more than 17,000 HMO members, researchers found that the more ACEs a child has, the greater the chances of long-term physical and behavioral health issues. ACEs even impact mortality; six or more ACEs drop a child's life expectancy by twenty years. More recently, research has found that just one ACE puts a child at a 28 percent increase in risk for asthma; four ACEs put her at a 73 percent increase. This new understanding of the health consequences of adverse experiences has changed how we practice medicine by broadening our field of vision—forcing us to see a child's total environment as *medical.* We aren't just looking at a child's physical condition on the day of an exam or clinic visit. We are looking for the larger factors in the child's world that can impede development and diminish an entire life—and may put her at risk as an adult for diabetes, heart disease, or substance abuse. This is the most important concept in pediatrics and public health today.

Science also shows us there's cause for hope. We may not be able to give every child a happy, healthy, and safe childhood—though we should keep trying. But we can mitigate the effects of adversity and toxic stress by building resilience. It's the key to development, the deciding factor between a child who learns to cope and thrive and one who never makes it to a healthy or productive adulthood. Resilience isn't something you have or don't have. It's learned. While the stress hormone response in a child overloads the child's system, it can reset to normal if she is soothed by caring adults in a nurturing, stable environment and community. The brain can heal.

It's important for my pediatric residents to read the most up-to-date literature and science from the leaders in the field about ACEs, toxic stress, neurodevelopment, and resilience. They watch tutorials on brain development and the impact of toxic stress from Harvard's

Center on the Developing Child and watch Nadine Burke Harris's TEDMED talk, a fan favorite in the curriculum. But it's important and much more galvanizing if they see it firsthand in the community where our children live.

So at the beginning of the Community Pediatrics block, residents go on a tour of the city and learn the history of Flint, from its days as a GM boomtown and the birthplace of union contracts and the middle class to its decades of dire decline. They record the number of blighted neighborhoods, liquor stores, neglected playgrounds, and boarded-up schools.

I can't assume all our residents know the history of racial injustice in this country, let alone the historic racism in medicine. So the curriculum includes webinars on race and health and a discussion of the story of Henrietta Lacks. (Everyone in medicine knows HeLa cells, but many of them don't know about their namesake, a woman whose life vividly illustrates medical racism and its consequences.) And our residents view *Unnatural Causes,* a seven-hour PBS documentary about socioeconomic and racial disparities in healthcare in America and their root causes.

We discuss the Tuskegee syphilis experiment, the infamous clinical trial that the U.S. Public Health Service conducted on six hundred African-American men between 1932 and 1972. Tuskegee participants were told that they were getting free healthcare for life, but in reality they were enrolling in a study of the natural progression of untreated syphilis. Even after the discovery and widespread use of penicillin, which cures syphilis, this inhumane experiment continued. The men—who were selected because they were poor sharecroppers with little education or recourse to the law—were still not treated. Untreated syphilis is gruesome, causing lasting damage to the body, with symptoms including nasty lesions and eye damage, as well as nervous system and cardiovascular breakdown.

Some of our white students, and some of our international ones—who make up a significant number of residents—find it hard to believe that this kind of racism ever existed or that it persists. But our African-American residents know different. Even if they don't know the specifics of these stories, they are all too familiar with the outlines of this ugly history and enduring reality.

Residents also meet with community leaders and activists, and they visit nonprofits and schools and daycare centers. They are sent to home visits, to court hearings and trials, to state protective services and community events. They meet Professor Rick Sadler, a recently hired MSU nutrition geographer and Flint history buff whose Flint "food desert" maps illustrate the role of nutrition access and food security in children's health. That summer Rick was helping me figure out if relocating the Hurley Children's Clinic to the space above the Flint Farmers' Market would help our patients improve their diets.

My objective for this Community Pediatrics rotation is to get the residents out of the hospital and into the city, into the lived experience and environment of our kids. They would become familiar with the city's weaknesses and needs but also feel a sense of solidarity and empathy with the people of Flint—and see the city's deeper potential. That's the feeling I wanted to imbue them with most of all: that there's hope in this town, not hopelessness. It just needs some nurturing and care to build. And they needed to see their privileged role as builders, shoulder to shoulder with our neighbors.

Flint has been through so much—after decades of downward spiraling, it has become beleaguered and almost bankrupt. But the spirit of the community never collapsed.

More than fifteen years ago, I fell in love with Flint as a medical student at Michigan State University. MSU, a pioneer land-grant university, founded the country's first community-based medical school in 1964 and reaffirmed its commitment to Flint in 2014, when it moved all public health programming to the city and expanded it. The medical school's motto is "Service to People." This sense of community investment and hope drew me back there when it was time to find a place to plant my own roots, and it drew others back too, like Dayne Walling, who at twenty-five was elected mayor in 2009, a couple of years before I returned to Flint.

I knew Walling by reputation. A Flint native, he had gone to

MSU's James Madison College, a political science residential college, and graduated a year ahead of my brother, Mark (who is now a public-interest lawyer in Washington, D.C.). Walling went on to study at Oxford on a Rhodes Scholarship and followed that with a master's degree in urban studies from the University of London. As soon as he got back to Flint, he got involved in local politics.

Flint seemed pretty lucky to have the leadership of someone so young, optimistic, and even telegenic. But things went downhill on the day of Walling's reelection in 2011, when Michigan governor Rick Snyder declared that nearly bankrupt Flint was in a state of "local government financial emergency" and appointed an unelected emergency manager (EM) to run the city, taking all real power away from Walling.

Snyder, a new governor, was popular at the time. He dubbed himself "one tough nerd" and had a history as a successful business executive. He was a Republican who ran as a moderate—a technocrat—but he was soon pushed to the right by a Tea Party–controlled state legislature. Flint wasn't the first Michigan city to have its democratically elected government replaced by an EM who demanded draconian budget cuts: Snyder had appointed EMs in Detroit and Pontiac. By 2013, half of all African-American citizens in Michigan were living under an EM, compared with 2 percent of white residents. In other words, half of the African-American population in Michigan did not have elected representatives running their cities—the cities had been effectively colonized by the state. This seemed grossly undemocratic to me and hardly an accident.

EMs didn't answer to the people. They answered to Snyder.

ONE OF THE BUDGET-CUTTING brainchildren of Snyder's emergency manager regime was to change the source of Flint's tap water. For half a century, Flint had bought safe, pretreated drinking water from the Detroit Water and Sewerage Department, a massive public utility that pumped water from Lake Huron and lucratively sold it to dozens of communities in southeastern Michigan. Tired of its "water dependency" and the steep prices charged by the Detroit utility, a team of elected officials from Flint and Genesee County—the county that contains Flint—along with members of the governor's office, decided that they should build a new parallel pipeline to Lake Huron. To save even more money, the state determined that until the pipeline to Lake Huron was finished, the stopgap water source would be the Flint River. This was the crucial mistake.

You didn't have to be a water expert to know what everyone in the area knew: the Flint River had been a toxic industrial dumping site for decades, even if in recent years the river water didn't look quite as brown or *as thick and flammable* (it was said to have twice caught fire) as it had before the 1972 Clean Water Act.

But was the water safe to drink?

That's why we have agencies like the Michigan Department of Environmental Quality (MDEQ) and the federal Environmental Protection Agency (EPA) in Washington—to answer that question. In my head, I pictured these agencies as populated by diligent bureaucrats in white lab coats with test tubes, studying water quality to ensure that what came out of our taps was safe. It was their job to protect our health and safety. And nothing is more fundamental to our health and safety than our water. I believed they took that responsibility seriously.

On April 25, 2014, I watched the news as Mayor Walling pressed a button at the Flint Water Treatment Plant to shut off the valve to close the Detroit water supply. When the pipes opened to the Flint River, he toasted by drinking a glass of the water. After that, I assumed that those people with their white lab coats and test tubes were doing their jobs—and that life in Flint would carry on with little difference. But almost immediately, complaints began to appear in the local media.

People said their tap water smelled bad and tasted worse. It was brownish. It was greenish. It was disgusting. The agencies did their testing. Soon the city released boil alerts because of bacteria, which didn't inspire confidence. So much chlorine was added to kill the bacteria that the tap water began irritating people's skin and

MAYOR DAYNE WALLING SWITCHING THE
WATER SOURCE, APRIL 2014

eyes. And it also led to a buildup of a disinfectant by-product called total trihalomethanes (TTHMs)—a carcinogen if inhaled. The city of Flint was found in violation of safe drinking water levels of TTHMs, but then that was apparently cleared up. The alerts ended and an all-clear announcement was made. The debut of the new water source wasn't flawless, but I had no reason to suspect that the agencies we'd entrusted to look after our water weren't doing their jobs. We were in America, not a developing country. It was the twenty-first century. And Flint was literally in the middle of the Great Lakes region, the largest source of freshwater in the world. Why doubt the safety of what was coming out of the tap?

What little I knew about drinking water I knew from my high school friend Elin Betanzo. After graduating from Kimball High as class valedictorian, she studied engineering and music performance at Carnegie Mellon, then drifted into the field of environment science, eventually clocking some mostly unhappy years in the EPA's Office of Ground Water and Drinking Water. I didn't remember all the details of her departure. And to be totally honest, when Elin brought up her work—with its details about "distribution systems" and "compliance" and "sampling variability"—sometimes I found myself mentally checking out. She was so in the weeds. And there were a lot of weeds in drinking water.

CHAPTER 2

The Barbecue

B Y THE TIME I GOT HOME THAT NIGHT, AUGUST 26, 2015, EL-liott had started dinner. Chicken had been smothered in some kind of super-secret barbecue rub. On the deck, his exotic egg-shaped Kamado grill was heating up and Elliott was struggling a little with its heavy lid. His right arm was in a sling.

Just a month before, he'd undergone shoulder surgery for various injuries that had compounded over the years from playing baseball and racquetball, then lifting weights. He was still in pain and probably shouldn't have been grilling at all. But Elliott loved his Kamado grill so much, it was hard to keep him away—sometimes we joked that it was the son he never had. He insisted he didn't need help, and truthfully I didn't know anything about barbecuing, so I rushed upstairs with just enough time to change into jeans and ask the girls about their time at Skull Island day camp.

Nina and Layla, nine and seven, my dark-haired, brown-skinned little girls, mostly get along, probably because they are so different. Nina is studious, quiet, and somewhat introverted and anxious. Layla, our baby, is direct and demanding—kind of like Mama Evelyn, a force of nature. You always know where things stand with her. They both loved their eight weeks of summer day camp but hated their first day of Skull Island. By the time I found them in the kitchen, they'd worked themselves into a frenzy of complaint. Most of their friends were away on family vacations, they said. Why couldn't we go away?

If inequality has a favorite season, it's summer. My brother, Mark, and I didn't grow up with luxuries like summer camp, and my patients in Flint rarely had chances to experience something that so many upper-middle-class families take for granted. Many of those kids are stuck at home, sometimes in neighborhoods where playing outside isn't a safe option. Academically, they tend to lose some of what they learned the previous year in school—what's called the "summer slide"—because they aren't kept intellectually stimulated over the summer. Meanwhile rich kids go to camp—and receive lots of academic and extracurricular enrichment, along with bonding time with their peers and healthy outdoor activities. Knowing about this disparity—and that we were on the privileged side of this inequality—kept my sympathies for Nina and Layla in check. Plus, I had too much work this time of year for a family vacation. It wasn't the end of the world for my girls, only the end of summer.

Annie and Elin arrived with their own kids and husbands, and we all began catching up. Our shared history was deep. In the drama club at Kimball High, we had been hams. In the environmental club, Students for Environmental Awareness (SEA), we were passionate activists, spending weeknights knocking on doors, circulating petitions, and organizing events. What we learned in those years had shaped us. Annie, an Irish-Italian beauty with long, straight blond hair and a tall, athletic build, was the founder of Opera on Tap, a network of opera singers who perform in bars, stores, and parks. Elin was now working remotely from Detroit for a D.C.–based think tank. We all had our own kids now, five little ones between us.

The last time the three of us had been together was my wedding—and we shared a few jokes about that. Wedding planning had been agonizing for a number of reasons but mostly due to my powerful aversion to the traditional wedding style of Detroit-area Chaldeans, the Iraqi Christian sect to which both Elliott and I belong. Chaldean weddings are Big Fat Greek Weddings on steroids—with *debke* dancing, Arabic pop music, professional belly dancers, and always a high-end open bar for up to a thousand guests. Hummus, baba ghanoush, and tabbouleh are followed by a full-course meal that doesn't start until 10 P.M., usually consisting of many types of kebab and saffron-infused rice. Leaving a wedding, or any kind of Iraqi party, without being absolutely stuffed is unheard of. Refusing to eat any of the multiple courses is a grave insult. For dessert, the yummiest weddings will have baklava from Shatila Bakery in Dearborn along with a traditional wedding cake.

But that kind of extravaganza didn't really fit with my lifestyle or worldview—or Elliott's. So we spent our minimal wedding-planning hours dreaming up ways to upend tradition. Which was how we came to throw our reception in the Detroit Science Center—we invited our guests to "Mona & Elliott's Science Project." And instead of "The Wedding March," we asked Elin, a pianist, and her husband, Mauricio, a cellist, to play the theme to *The Princess Bride* when I walked down the aisle.

And rather than the usual two dozen bridesmaids and groomsmen, our small "wedding party" consisted entirely of immediate family members. My brother, Mark, and his wife, Annette, were my "maids of honor." Elliott asked his sister, Angie, and brother, John, to be "best men." The whole thing turned out to be a wonderful and fun experiment. Except I will never live down setting the church on fire.

Annie was singing Schubert's "Ave Maria," her hypnotic soprano voice flowing through the church. Elliott and I were standing together, looking at Father Frank on the altar. We were calm and taking everything in stride. I was a young pediatric resident beginning my life with another pediatrician, a creative and sensitive guy. Before we met, I had had three rules of dating: (1) no doctors, (2) no Chaldeans, and (3) no Republicans. Elliott was 2.5 of the 3. (Raised in a conservative family, he admits that he once voted for a Republican.) But the force of his charm and romantic persistence won me over.

The ceremony itself was straightforward. And now before us, on an altar cloth, were two small candles, both lit, representing Elliott and me. In the middle was a large candle that wasn't lit. As Father Frank had instructed during the rehearsal, we were each supposed to take a small candle and together light the big one. This act would symbolize our union.

The week before, I had successfully performed my first spinal tap on a newborn. Light a candle? Piece of cake.

We lit the big candle in the middle of the table without incident. Then, as I tried to make sure my little candle was safely returned to its holder, I must have pressed down on it too hard. (I'm not known for my light touch; my handshakes are notoriously firm.) The candle toppled over. Within seconds, the altar cloth was in flames. Elliott reflexively tried to put the fire out with his hand and burned his fingers, while Annie's singing of "Ave Maria" seemed to grow louder, fueled by our little inferno.

Flames were leaping. Smoke was now rising around us. Then I heard Mark's deep voice bellowing behind me—he couldn't stop

JUST BEFORE THE ALTAR CAUGHT ON FIRE,
FATHER FRANK LOOKING ON

laughing. Others in the pews began to chuckle. I turned around to see the face of my mom: sheer horror.

Amid the smoke and Mark's infectious laughter, Father Frank frantically smothered the flames. Elliott and I just looked at each other, smiling; he grinned through the pain of singed fingers. Elin, Annie, and Mauricio played on, flawlessly, in perfect harmony, as if nothing were happening.

We wanted an unconventional wedding, and we got one.

So here we were—more than ten years later—standing around the flames of the smoky Kamado grill while Elliott cooked the chicken. To honor the reunion of the survivors of our wedding fire, we opened a bottle of wine that Elliott and I had brought back from our sort-of honeymoon ten years before. I say *sort-of* because even back then, I couldn't fully unplug from work mode. I combined a honeymoon with a pediatric residency rotation at the American University of Beirut Medical Center.

The wine was Ksara, from the famous vineyard in Lebanon. We had saved it for a special occasion. Not being wine connoisseurs, though, we didn't know about its short shelf life. After a decade, it was undrinkable.

We opened another bottle and watched our kids form an impromptu rock band and crank the Block Rocker, their karaoke machine, up to its loudest setting. The toy guitars and plastic drum set were all super cute, but the noise sent me to the corner of the kitchen, where Elin and Annie and I caught up, wineglasses in hand.

"How's work going?" Elin asked, zeroing in on me for some reason.

"Things are awesome," I said, which is what I always say, being an innate optimist, along with *super!* or *amazing!* even if things aren't. I allowed that my job at Hurley was demanding but becoming a little routine, but it's not like I had the time to do more. Already there weren't enough hours in the day.

Annie went next. Her life wasn't any calmer. She had a young daughter at home and no family nearby to help, and her nonprofit was exploding worldwide, while she was also running her own Web services and business consulting company.

Then it was Elin's turn. She was proud of her Swedish name and background, but to me—a dark-skinned, dark-haired immigrant kid from the Middle East who had garlicky hummus in my school lunches—Elin always seemed like the quintessential all-American girl. I'd envied her fair complexion, gray-green eyes, and light brown curly hair when I was younger, along with her other gifts. In the Advanced Placement classes we took together at Kimball, Elin had sailed along, getting perfect scores and grades without breaking a sweat and excelling at the piano at the same time. It's funny how strong these high school impressions can be, because even though I knew from our recent visits that Elin's career had hit the doldrums, I continued to think of her as one step ahead.

Always precise with her words, she confessed that she'd relocated from Washington to Michigan to be closer to her family, but working from home felt isolating. She didn't sound that enthusiastic about her projects either. Life had forced too many compromises on her. Maybe being the valedictorian came with unsustainably high expectations.

Then she turned to me with a total non sequitur. "What are you hearing about the Flint water?"

The question surprised me, as did the abrupt shift in Elin's demeanor. Her eyes were suddenly boring into mine.

"I've heard the complaints," I said. "But the state says it's fine—in compliance."

"No," she said, shaking her head. "It's really not."

CHAPTER 3

The Valedictorian

ELLIOTT AND I HAD BEEN INDOCTRINATED BY OUR PARENTS: being an Iraqi host means offering your guests more food than three times as many people could eat. That night was no different. Elliott finally pulled the slightly charred chicken off the grill, and our noisy rock band took a break and wandered toward the heaping platters of food.

As soon as we got the kids settled, cutting up their food into kid-size pieces, Elin and I found each other again. She talked quickly, with intensity and passion.

She had seen a memo written by a former colleague of hers, Miguel Del Toral, who worked in the Chicago office of the EPA. He had come to Flint a few months before, after being contacted directly by a resident who was concerned about her water. He personally arranged to have an independent test of the tap water in the resident's home.

"I worked with Miguel. I know Miguel," Elin said. "I trust him. He wouldn't write this memo if there wasn't a serious problem."

"And?"

"He says that Flint is not using corrosion control."

The urgency on Elin's face was unmistakable, but I was still a little lost.

"What is corrosion control?"

A shriek rose up in the next room. The posse of kids was already restless and returning to the drum set, guitars, keyboards, and karaoke.

"A system of Flint's size is required to have corrosion control as part of the Safe Drinking Water Act. Water is naturally corrosive, but water systems are supposed to treat the water to reduce corrosivity. When you change the source of the water or how it's treated, this changes the way the water reacts with the pipes—that's what we learned from the D.C. water crisis."

The D.C. water crisis? I didn't think I'd heard of that. The music began again, worse and even louder. Elin and I tucked into a deeper corner of my kitchen. "Whether corrosion is a problem or not depends on what the pipes and plumbing are made of," Elin went on. "In Flint, the service lines, the pipes that run from the street into people's homes, are made of lead. And the plumbing inside people's homes has lead too."

"Wait a second," I said. "Are you saying—"

Elin nodded. "If Miguel's right that Flint is not using corrosion control, that means there's lead in Flint's water."

"Lead in the water?"

Elin waited a few moments while I processed the news, then began nodding slowly. "And based on Miguel's memo," she went on, "the lead levels in the Flint water are really, really high. He suspects that MDEQ isn't testing correctly. That's why he leaked the memo."

"Are you kidding me?" I shook my head. "Why would anybody at the EPA need to leak their own memo?"

Elin cocked her head and just stared at me, deadpan. She was waiting for me to catch up.

"Okay," I said, moving on. "So how could the testing not be done right?"

"MDEQ is probably testing for the results they want, which will underestimate the amount of lead in the drinking water. The loopholes in the Lead and Copper Rule allow for that. The utilities do it all the time—trying to game the regulations and manipulate the data to minimize the amount of lead collected in a sample. I don't trust them."

I am not naïve. I was only twelve when I saw photographs of children who had been poisoned by Saddam Hussein, when he gassed the Kurdish town of Halabja and murdered thousands of his own people. I had nightmares for weeks. I learned then what governments were capable of.

But here in Michigan?

It was one of those moments when part of your brain whispers, *Please, can we go back to five seconds ago, before we knew this?*

If what Elin was saying was true, it meant my kids, my Flint kids, who were already struggling with so much adversity, were in even greater danger than I'd imagined.

I know lead. All pediatricians know lead. It's a powerful, well-studied neurotoxin that disrupts brain development. There is truly no safe level.

What about my imagined government scientists with their white lab coats and test tubes? What had happened to them? They were supposed to make sure our water was safe.

So far only one lone bureaucrat seemed to be doing the right thing and he had to leak the information to get it out. But if Miguel Del Toral was right, an entire city was being lied to. And the liars were the very people we trusted to keep us safe.

I thought about all the complaints about water in Flint over the past year—the state had answered them with a parade of assurances. Just five months before, because of general water-quality issues and an outcry from residents, the Flint City Council had voted to return to Detroit water, but Governor Snyder's appointed emergency manager described that decision as "incomprehensible" and rejected it as too costly.

How could money be more important than clean water? Or the safety of our kids?

And just a month ago, the mayor himself had gone on live TV, filled another glass with Flint water for the cameras, smiled, and drunk it.

I DON'T KNOW IF I'd ever felt so stunned and disillusioned and sad all at once. What had I done? Where had I been?

Baby after baby had come into our clinic. We gave the same advice. *It's fine. Yes, it's fine. Drink the water. Of course it's okay.* Flint has low breastfeeding rates for a number of reasons. Powdered formula is the norm. To make formula from the powder, you have to mix it with tap water. Meanwhile, for older kids, because of our emphasis on healthy living and lowering sugar in their diets, we are always recommending fewer juices and soda. And more water.

It's fine. Drink the water. Of course it's okay.

When my pediatric residents came to me with questions from patients, I said the same thing, just as I had been taught: *Tap water is the best and safest.*

I drink it. The coffee I mainlined all day was made with it.

Flint water.

Lead.

I thought of Reeva and her sweet baby sister, Nakala, who had started the shift to powdered formula the week before. How many days had passed? Four or five days of tap water *with lead.*

Little Nakala wouldn't feel any different. The poisoning would be quiet. Lead exposure is known as a silent epidemic because there are no immediate signs of it. But once it was in her bloodstream, the lead would enter her red blood cells and wreak havoc, interfering with the mitochondria, the part of the cell where energy is produced. It would go on to disrupt the formation of the dendrites, disturb the myelin sheath that surrounds nerve fibers, and interrupt the way hemoglobin carries oxygen through Nakala's body.

It would settle in her soft tissue and her bones, where it would crowd out calcium. Some heavy metals—iron, copper, zinc, selenium—have health benefits in small amounts. But not lead. It does only harm.

Nakala's central nervous system would receive the worst impact from the lead, primarily in her still-developing brain. A significant amount of lead exposure can cause swelling of the brain, headaches, lethargy, anemia, dizziness, muscular paralysis, sleeplessness, loss of appetite, and abdominal pain. It can affect vision and hearing, and in extreme cases, it can cause kidney failure, coma, and death.

Even at lower, single-digit levels, the damage is irreversible. For an infant like Nakala or a two-year-old like Reeva, enough exposure can mean developmental delays, cognitive impairment—literally, a drop in IQ—as well as memory issues, attention and mood disorders, and aggressive behavior.

Brain scans show that lead exposure in children causes an erosion of gray matter that makes it harder to pay attention, regulate emotions, and control impulses. It also affects the white matter of a child's brain, which acts as a conduit for signals within the central nervous system. As lead-poisoned kids reach their teens, they have a much harder time in school and are more likely to drop out. As they reach their twenties, research suggests, they may be more likely to commit violent crime. Lead is even suspected to have an epigenetic or multigenerational impact by changing a child's DNA. It's really science-fiction comic-book stuff, like the X-Men, except the victims aren't getting superpowers. Their powers are being taken away.

HUDDLED IN OUR CORNER of the kitchen, Elin and I continued to talk about the leaked EPA memo. It felt as if a two-person pup tent had dropped over us, and the rest of the party had gone away.

"I need more wine," I said, and hunted for the open bottle.

Lead in the water. Now that I knew the truth, I couldn't un-know it. I could only go forward.

"Let's think for a second."

"I don't want to sound pessimistic," Elin said, "but you're probably looking at years of fighting this before anything will be done."

It was Elin's nature to be cautious and circumspect. When we met, I had been young for ninth grade, only thirteen, an extrovert who went out for every sport, every activity, every play, and every musical, even though I had limited athletic talent and a pretty terrible voice. Elin was almost a year older, an introvert with an engineer's mind who looked at life as a series of problems and puzzles to solve, slowly and carefully. She was the kid who read every page of her assigned work—and then read all the books that it referenced. By our senior year of high school, I was class president, and she was class valedictorian. And while I was perpetually upbeat and talkative, my old friend was, well, kind of ruminative and quiet.

Her career had emphasized this even more. Her degrees in environmental science and environmental engineering had led her to work at government bureaucracies where she was always rolling a giant boulder uphill, going against improbable odds, doing endless paperwork exercises, and struggling against political tides. Results came after decades, not after days or months—something that would have been torture for an impatient person like me. We were both newly married and starting families in the years when she was working at the EPA. She had been scarred by that experience, but I didn't really know how.

"I had a bad feeling about Flint from the beginning," she went on, describing a water-quality meeting she had attended in Michigan the previous year, when she first heard about the water switch. It was supposed to save Flint money and to be better down the road, or at least that's how it had been explained at the time. "It didn't make any financial or engineering sense to change from Lake Huron. Water source changes are very rare. *No one goes from a high-quality water source to a lower-quality source unless they are running out of water.* And Lake Huron wasn't going anywhere. The infrastructure was all in place. But there was a new requirement that states do an evaluation to make sure that corrosion isn't a factor when a water source change occurs. So I reassured myself with that."

Elin's and my longest and deepest bond was environmental activism, something we also shared with my brother. We had been profoundly influenced by Roberta Magid—or "Bobby," as we secretly nicknamed her—a feisty, progressive New Yorker and high school librarian who was the faculty adviser to our high school environmental club, SEA. But SEA was more than a school club. It was a serious real-world undertaking where we learned to strategize and create real change through direct action. Sure, we did the regular stuff of any high school environmental club, like recycling cans and putting on eco-plays for elementary school kids. Mark played an evil Trash Monster, Elin was Aqua Anastasia, and I was always Mother Earth.

But Roberta encouraged us to take it further—to take risks, to put ourselves on the line. We studied activism by becoming activists—going beyond changes in individual behavior to explore how policy and politics helped or hurt our causes.

AN ENVIRONMENTAL FAIR WITH ELIN, NINTH GRADE, 1991

———

OUR BIGGEST UNDERTAKING WAS the Madison Heights incinerator. An elementary school stood in the shadows of its smokestacks, and from soil samples and air-quality reports commissioned by a local community group, Clean Air Please (CAP), we learned that the adjacent neighborhood had higher rates of asthma and chronic obstructive pulmonary disease. For decades, soot from the incinerator coated these modest working-class homes. The county closed the incinerator for a time, but when it tried to reopen it, we went to the streets with CAP, led by a courageous nurse and single mom, Pam Ortner. A group of us—including Elin, Annie, Mark, and our classmate Dave Woodward—protested, knocked on doors, and wrote letters.

Then we campaigned to elect John Freeman, an environmentalist and lawyer, to the state legislature. Once in office, he added a provision to a state law to make it impossible for an incinerator to operate so close to a school. That was the end of the Madison Heights incinerator. It closed down for good in 1992, our sophomore year of high school.

Standing there in the kitchen, I thought back to Roberta and Pam and John and their leadership. How much they had inspired me. How committed they were, and how much they cared. Their activism combined environment, health, and policy at once—and made the world a little better.

Once you have had a chance to change things, to have a real-world impact, you never forget it. I felt myself tapping back into that powerful feeling, but the urgency and scope of the problem felt overwhelming.

"*Years?* But how could this take years to resolve?" I asked, incredulous. "Lead is the worst kind of poison. Permanent. Life-altering. These kids can't afford another day." I thought Elin was being too pessimistic.

"Remember what I went through in D.C.?" she asked.

I shook my head. I was embarrassed not to remember an important story from Elin's life that she apparently had told me, but I didn't have time for apologies at that moment. "Remind me."

"Remember when I left the EPA?" she went on. "And why I refused to work for D.C. Water?"

I searched my memory, trying to jar something loose. "There was somebody you were trying to stay away from, right?"

"Something like that. Not only did nobody get punished for the D.C. crisis, some were promoted. What you need to know tonight is that, basically, scientists and activists tried to prove that children were harmed, but they couldn't get the health data to show it. Lead was in the D.C. water for years. More lead than you could imagine. More lead than I want to think about."

This shocked me. Where had I been?

"D.C. is the worst-case scenario," Elin said. "They made a treatment change but didn't study its effect on water quality in advance. When they started measuring high lead in the water, they just kept sampling to try to bring down the system average, but in fact they kept measuring lead at higher and higher concentrations. And they didn't tell the public. Nothing happened until *The Washington Post* put the story on the front page three years after it started. Even then, the government and utility demanded proof of impact, but never really got it."

"Proof of impact, like—"

"Proof. Like, researchers couldn't draw the link between increased lead levels in water and impact on kids. No link, no traction. No link, no action. Just a lot of stalling and—"

She stopped in midthought. "Wait, aren't you in charge of the hospital in Flint? Or something like—"

"No, I'm in charge of the pediatric residents, and—"

"But don't you have access to the kids' blood-lead tests?"

"Yes. Of course I do."

Kids on Medicaid are supposed to have their blood-lead levels tested at one and two years of age. My mind began working so quickly, thinking about how to get hold of the blood data that I'd need, that I forgot to keep up my end of the conversation.

"Mona."

"Yeah." I wasn't really hearing her.

"Mona, listen to me."

"What?"

Elin moved very close to me, inches away. Her gray-green eyes were emitting a laser sharpness and energy that would have been scary if I hadn't known her so well.

"You can do this," she said.

"Okay."

A fog of unreality fell over the rest of the party. Elin's words pinged and echoed in my head as the ragtag rock band devolved into chaos, and two of the littlest kids had very dramatic meltdowns. There were cries and pleas, *Mommy, Mommy!* It was late, past their bedtimes. We moved our guests quickly to the front door, where I gave everybody a hug and a short goodbye. With difficulty, Elliott put his sling back on. We cleaned up, did the dishes, and went upstairs to put the girls to bed.

—

Haji

I WAS DISTRACTED, MY HEAD STILL SPINNING FROM THE CON-versation with Elin. All I could think about was getting to a computer and searching for papers on lead exposure, on the D.C. water crisis, on water treatment chemicals and their corrosion effects. But Nina and Layla were asking for a bedtime story. They wanted me to tell them "Haji and the Birds." They usually heard it from my mom or my brother.

"Oh, I'm not good at telling that story," I said.

"Come on," Nina said softly.

"Tell us!" Layla insisted.

"What if I just talk to you about Haji instead?" I said. "I don't think you remember him."

"I do," Nina said.

Elliott entered the room. "Are you telling Haji stories?"

"Yes!" The girls called out.

"Okay," I said, sitting on the edge of the bed with Elliott, all four of us together. "Just like you guys are learning a lot of amazing stuff from your grandparents, like how to speak *Arabee*, make yummy food, and grow a garden, I learned lots of things from Haji."

Layla was starting to drift off, but fighting it.

Nina, a night owl, was hanging on every word.

"Haji lived far away, in Baghdad, where he had many successes, many jobs and businesses, and where he had so many friends. They were all kinds of people with different religions and different ethnic backgrounds—Kurds and Jews and Muslims. He was an idealist. Do you remember what that means? It's somebody who believes the world can be better than it is now. Haji had great faith in people but less faith in religion."

"Did he go to church?" Layla asked. I figured she was curious because, unlike most Chaldean families, we weren't regular church-goers.

"No, he didn't."

"Why not?"

"There's actually a story about that," I said.

"Tell us that story, Mama."

"Okay."

Layla smiled and settled deeper into her bed.

"Haji had four brothers. All their names ended in an *L*. I don't know why. And Haji's real name was Khalil, so his name ended in an *L* too. He had two sisters. Do you remember their names?"

"One of them was Layla!" Layla said.

"That's right. They were all very creative and started many businesses. One uncle was a lawyer. Another uncle opened a famous bookstore called Dar Al Katub Al Kadema, which means 'The

House of Old Books.' It was the meeting place for idealists in Baghdad. Another uncle was a prominent chemist who graduated from American University in Beirut, and another became a lawyer. And Haji was a successful businessman who traveled around the world and later became director of Pepsi-Cola in Iraq.

"But at one time all the brothers were in business together. They owned a perfume factory. Haji enjoyed tinkering with various oils and scents to create amazing, exotic fragrances. He had an amazing nose," I said, touching the tip of Layla's. "Later, they made plastic prayer beads and sold them to Shiite pilgrims traveling to the holy cities of Karbala and Najaf.

"We have a lot of chemists and scientists in our family, don't we?"

The girls nodded.

"Because of their reputation for mixing chemicals, during World War II, the British Royal Air Force asked Haji and his brothers for advice. It was so dry and windy in the desert of Iraq, and there were so many sandstorms, that the paint on the British airplanes was peeling off. After some experimentation, Haji and his brothers melted reels of old movie film and mixed it with the paint, and the problem was solved. In addition to keeping the paint from peeling, the brothers' paint recipe seemed to make the planes invisible from the ground—and harder to shoot down.

"Pretty cool, huh?" The girls were captivated.

"Around that time, Haji's perfume factory burned down. He and his brothers lost everything in this fire. As the factory smoldered, or burned down to ash, they sat on the corner across the street. They were inconsolable. Do you know what that means?"

"Keep going," Nina said, nodding.

But I knew she didn't know what *inconsolable* meant.

"They were feeling as sad as they'd ever felt, and unlucky. Nothing could lift their mood. While they were sitting on the corner, a long line of people came to say how sorry they were. All their friends came, their neighbors, the local imams, and a rabbi. Finally all their employees came, to offer words of support and encouragement. *Don't give up! Just build a new factory!*

"Haji said that wasn't possible. He didn't have the energy, and he was feeling too defeated. But the employees kept encouraging him. *Don't give up!* When Haji said he didn't have the money to build a new factory, the employees offered him money. *How much do you need?* they asked. One by one, they began handing him Iraqi dinars.

"This impressed Haji so much. The employees had little money of their own, but they wanted to help. Afterward, whenever he told the story of his perfume factory burning down, he would cry. He was moved by the goodness of his friends and employees. Their generosity consoled him—made him feel better. It helped him to be brave and rebuild. And he was no longer sad."

Layla looked me sideways. "Ha!"

"But he noticed that none of the Iraqi priests offered him help or support—or even a kind word. One priest told Haji that the fire was God's punishment for his arrogance. He said that Haji believed in people more than in God. Haji stopped going to church after that. He said he would never step in a church again."

"Never?" Nina asked.

"Almost never," I said. "But you know, what the priest said was true. Haji *did* believe in people more than anything else. And they believed in him. He was a humanist, someone who believes that people can make the world better. Haji respected everyone's beliefs and never spoke against any religion. But he didn't believe that God wanted calamity for anyone—or that anyone deserved to be abandoned to fate or bad luck. He taught me to treat everybody well, because we are all equal, no matter what we look like, what we believe in, or how much money we have. To always do the right thing, even if it's hard. Even if people tell you it's impossible. And maybe that's even better than going to church."

I looked over at Elliott, who was now smiling too, remembering Haji.

"Even today," I went on, "kids and grandkids of Haji's employees reach out to us, totally out of the blue, to tell their Haji stories, and they always mention his generosity. So their gratitude is passed down through stories and extends to us, even to you, Haji's great-granddaughters. Because Haji's actions in life, and things he believed in, left a mark on us all—even though we can't see it."

—

CHAPTER 5

Red Flags

AFTER THE GIRLS FINALLY FELL ASLEEP, HOURS PASSED, A BLUR of time in the dark with only the brightness of my iPad screen. As soon as Elin arrived home and put her kids to bed, she had sent me a long email with links to Wikipedia pages, Centers for Disease Control (CDC) reports, and EPA memos. I was up for three or four hours in bed, deluged by reports about water quality, water testing, water treatments, water pipe corrosion, and lead. Totally consumed, I forgot to sleep.

From time to time, I heard Elliott shifting. He was restless, trying to sleep in a big recliner that we'd borrowed from my parents and hauled up to the corner of our bedroom. The barbecuing had probably been too much for his shoulder. He was in pain and hooked up to an ice machine. Usually the continuous vibratory hum and pumping sounds were hypnotic and put me to sleep, but not tonight.

He asked me what Elin and I had been so intensely talking about. I thought he might have overheard something, but I didn't want to get into it. I said quickly, "Oh—nothing, nothing." He knew me well enough to let it go.

FROM: Elin Betanzo
TO: Mona Hanna-Attisha

SENT: August 26, 2015, 11:57:56 P.M.
SUBJECT: Lead in drinking water and Flint

Hi Mona,

Here is a lot of information about what I understand about the drinking water in Flint and a comparison to what happened in Washington, D.C. I'd be happy to help you wade through it if you'd like. If you know of any ways that I could help out in Flint let me know.

As the hours passed, I was stunned by how Elin had anticipated the water problems before they'd begun. The year before, when she learned about the water switch, she had offered her expertise as an independent water-quality engineer to help ensure that the new water supply was safe, but she had been turned down, told she wasn't needed. She was told everything was fine.

I started my reading with a few stories by Ron Fonger, an intrepid reporter at *The Flint Journal* and MLive.com, its online platform. Fonger had done a solid job covering the water switch—the residents' concerns and their activism, the parade of assurances from the city and state—but it felt as if nobody had been listening or paying attention. Not the people in power, anyway. I moved on to a series of articles by a Detroit investigative reporter, Curt Guyette, who had been assigned to cover the emergency manager story for the Michigan chapter of the ACLU, with funding from the Ford Foundation. Their concern was that citizens' rights to self-government might be usurped by a state-appointed EM. The collaboration between the ACLU and the Ford Foundation was a first for both of them, a philanthropic effort to fill the gap in costly investigative work that traditional media could no longer afford.

Guyette had been a skeptic about the water source switch all along. He dogged officials with questions they never answered, or at least not to his satisfaction. He tried to dig deeper into the story but didn't really find answers until the EPA memo was leaked to him.

In July 2015 he posted a news story on the ACLU blog and *Deadline Detroit*, releasing the memo to the public.

The memo—an eight-page interim report written by Miguel Del Toral, Elin's former colleague and the regulations manager of the Midwest water division of the EPA—was methodically detailed, science-driven, and shocking.

The words "high lead levels in Flint" were right there at the top, in the subject line under the official EPA letterhead. Honestly, that alone should have been enough to stop everyone in their tracks. It should have prompted the state to study and fix the situation and, most important, alert the public. You don't mess around with lead or even the suspicion of lead. Any pediatrician will tell you the same thing. The split second this memo saw the light of day, a state of emergency should have been called. And the governor should have asked for federal help.

But somehow here I was in bed, my hair on fire, fully two months later, reading about it for the first time.

Del Toral's report described water testing in the home of Lee-Anne Walters, a thirty-seven-year-old mother of four and military spouse who lived on Browning Avenue, on Flint's south side. Just a few months after the water source switch in 2014, Walters noticed that her three-year-old twins, Gavin and Garrett, broke out in red bumps after they were given a bath. Gavin had immune deficiencies and was especially prone to problems. If he soaked in the bathtub for a long time, a scaly rash would form across his chest at the waterline. As Curt Guyette's stories described, even after the city lifted its boil alerts and officially announced that the bacterial content in the water was in compliance with federal guidelines, the Walters family continued to experience rashes, along with abdominal pain and bizarre hair loss.

LeeAnne's eyelashes were disappearing. Her eighteen-year-old daughter, Kaylie, was taking a shower when a clump of hair fell out in her hand. And even more troubling, according to a Guyette story, Gavin had stopped growing.

Nine months after the water switch, in January 2015, Walters

UNITED STATES ENVIRONMENTAL PROTECTION AGENCY
REGION 5
77 WEST JACKSON BOULEVARD
CHICAGO, IL 60604-3590

REPLY TO THE ATTENTION OF:

WG-15J

June 24, 2015

MEMORANDUM

SUBJECT: High Lead Levels in Flint, Michigan – Interim Report

FROM: Miguel A. Del Toral
            Regulations Manager, Ground Water and Drinking Water Branch

TO: Thomas Poy
            Chief, Ground Water and Drinking Water Branch

The purpose of this interim report is to summarize the available information regarding activities conducted to date in response to high lead levels in drinking water reported by a resident in the City of Flint, Michigan. The final report will be submitted once

---

went to her first city council meeting. The following month a city employee was sent to test the water in her house. Prior to taking a sample, the employee ran the tap for a few minutes—explaining that he was following MDEQ guidelines to "pre-flush." This struck Walters as strange—if there was something in the water to worry about, wouldn't whatever it was get flushed out too? People don't usually flush tap water before they drink it.

Even with flushing, the sample came back showing an astronomically high amount of lead in Walters's water. Particles and metals in water are measured in parts per billion, or ppb, and the lead in Walters's water tested at nearly 400 ppb. Based on the latest scientific studies, the maximum amount of lead that should be allowable in water is 0 ppb. Even the EPA agrees. But the official "action level"—the amount of lead that warrants some sort of action be taken—is set at 15 ppb.

The memo stated that Walters took her children to the doctor for testing that revealed elevated lead levels in their blood.

When Walters notified the city of her children's high blood-lead levels, she was told that her household plumbing must be to blame, since the tests of city water showed lead levels in compliance with regulations. The city arranged for a garden hose to run from a neighbor's house to Walters's house to provide "lead-free" water.

A few days later Governor Snyder's office issued a statement that the Flint system was "producing water that met all the state and federal standards."

Walters was not reassured. For starters, when she first moved into her house, she had discovered that all the copper pipes had been stripped and stolen (not uncommon in Flint). That had forced her to install all-new plumbing made of polyvinyl chloride (PVC), a strong but lightweight plastic. She had told this to the city employee when he came to test. If all her pipes were made of plastic, how could they be leaching lead? Frustrated, she contacted the regional EPA headquarters in Chicago and eventually reached Miguel Del Toral.

According to Elin, only a handful of people in the drinking-water universe were as careful, and knew as much, as Del Toral. During her years at the EPA, where she wrote new regulations for the Safe Drinking Water Act, she worked closely with him on a "disinfection by-products" work group. As a member of that work group, Del Toral had a reputation for high standards, extra vigilance, and immunity to the pressures of groupthink. He raised concerns about drinking water regulations that everybody else dismissed or overlooked. Finally, here he was! My idealized lab-coat-wearing public health civil servant come to life. He didn't mind conflict and was unafraid to go his own way.

By the time Walters spoke to Del Toral on the phone, she had searched the Web to educate herself thoroughly about water treatment and had come to some conclusions. She told Del Toral she suspected the Flint water supply was corrosive and the pipes were leaching lead. She questioned whether the city was using corrosion control.

Del Toral found it hard to believe that the city wasn't using corrosion control and asked Walters to read him the list of chemicals that were used to treat Flint's water, as posted on the city website. When she was finished, he asked her to read him the list again, just to make sure he'd heard correctly. Not using corrosion control was a breach of federal law. There were no corrosion control chemicals on the city list.

He followed up with MDEQ immediately, but his calls and emails didn't get him anywhere. No action was taken to correct the problem.

Next, in April 2015, Del Toral traveled from Chicago to Flint to investigate Walters's house himself. First he thoroughly examined the plumbing, the pipes the city officials had described as the likely source of lead.

As his memo states, his inspection of LeeAnne Walters's home found that all the interior plumbing, including the pipes, valves, and connectors, was made of plastic certified by the National Sanitation Foundation (NSF) for use in drinking water applications. Del Toral's sampling showed that the faucets were also not the source of the high lead levels.

On his second visit, in May 2015, Del Toral deduced that the problems weren't inside the house but in a service line that led to it, when he extracted portions of the line and discovered it was made of galvanized iron and lead. Del Toral also found the pre-flushing procedure done at Walters's house prior to sampling very troubling. Pre-flushing was known to result in a "minimization of lead capture and significant underestimation of lead levels in the drinking water," according to his memo. This was a "serious concern" because the results "could provide a false sense of security to the residents of Flint regarding lead levels in the water and may result in residents not taking necessary precautions to protect their families from lead in the drinking water." But when he raised his concern with MDEQ, the department pushed back. Pre-flushing was not specifically prohibited by federal law, or by the Lead and Copper Rule, which governed drinking water. The department said it planned to continue the practice.

His conclusion: the pipes serving the Walters home were leaching lead. MDEQ was using faulty testing procedures. And Flint wasn't using corrosion control. He included the blood-lead levels of the four Walters children as evidence of impact, and he predicted that Flint could be facing an epic water crisis. He recommended that the EPA offer the city and MDEQ technical assistance.

He directed his report to Thomas Poy, chief of the Ground Water and Drinking Water Division of the EPA, but seven other individuals were copied, including two other employees at the EPA and four officials at MDEQ. The report was dated June 24, 2015.

Something truly disturbing happened after that. Rather than generating an instant response to the potential crisis outlined, Del Toral's findings were second-guessed. Rather than seriously investigating the matter as quickly as possible, as any responsible federal agency would, the EPA stalled, then reprimanded Del Toral for overstepping his responsibilities and referred him to the agency's ethics office. At MDEQ, the memo was met with even more resistance and hostility. State regulators painted Del Toral as a lone wolf and alarmist—"a rogue employee"—who had grossly exceeded his authority. Embarrassed, the regional manager at the EPA even apologized to MDEQ for Del Toral's memo, describing it as premature and incomplete.

Seeing no alternative, Del Toral, still defiant, gave Walters a copy of his findings, which she passed along to Curt Guyette, who broke the story on July 9, 2015.

ELIN AND I EXCHANGED texts and emails while I read. Sometimes she answered my questions in quips and scientific shorthand, but mostly we expressed ourselves in words of disbelief and angry emojis. The more I knew, the angrier I became. I was mad at myself for not paying attention to the repeated outcries of the Flint activists and even angrier at all levels of government for their lack of response and their gross indifference.

Red flags, so many red flags—they were everywhere. The information was all right there in Del Toral's memo. But somehow nobody with the power to make a difference had cared enough to notice.

LOOKING BACK OVER THE summer, I saw how I had been caught in my own busy bubble—dealing with a new crop of pediatric residents on July 1 and Elliott's shoulder surgery in mid-July, followed by the opening of our brand-new Hurley Children's Clinic, which we celebrated with an open house on August 7. We began receiving patients just a few days later. To make the summer more complicated, Layla's birthday was August 14, and I planned her party and the annual Hurley Pediatric beach party for the same summer weekend.

These are my excuses. But the truth was, after a year of concern about Flint's water being brown and smelly, and a year of assurances from officials, I'd failed to notice the dust storm caused by Guyette's pieces and had only a faint memory of government assurances about the water and Mayor Walling drinking a glass of it on television. But while I was consumed by the day-to-day distractions of my family and work projects, other people in Flint were speaking out—moms, activists, pastors, and even our kids. They were organized and vocal. Why wasn't anyone in government listening to them? Was it because they were poor and predominantly black? Or was it because there was no reason for the government to listen to the people because Flint was no longer a democracy? The city was in the hands of an unelected EM who was accountable to the governor, not to the people of the city.

I quickly ran the numbers in my head. Flint had about ten thousand children under the age of six. Many of them were our patients. Over the last year, we had treated thousands of Flint kids for winter coughs and summer sniffles, for strep, broken bones, and fevers. We weighed them all, checked their ears, and listened to their heartbeats carefully. We asked their moms if they were getting proper nutrition. We discussed obesity, drugs, and violence in the community. We gave them immunizations and, when they needed help navigating the system, referrals to social workers. We tried to help them get to their appointments with transportation assistance.

Being a pediatrician—perhaps more than any other kind of doctor—means being an advocate for your patient. It means using your voice to speak up for kids. We are charged with the duty of keeping these kids healthy.

We took an oath.

Where had we been?

Where had I been?

Now I was wondering whether I'd ever sleep or eat again.

"If you want to get really motivated," Elin wrote in an email, "look at the Wikipedia page on the D.C. crisis."

"Okay."

"You'll see what you're up against."

WASHINGTON, D.C., IS A mecca for recent college graduates. Entry-level jobs and internships in politics, nonprofits, and government are plentiful—and promotions can happen quickly. Elin's career was a vertical blur, as far as I could tell. And it was because Elin was in D.C. that we had stayed in close touch over the years. She lived only a few houses away from my brother in the crunchy suburb of Takoma Park.

Her first job, right after graduating from Carnegie Mellon, was at the U.S. Patent Office, but after a year, she found a job at the EPA, where she really wanted to be. The concept of environmental protection really meant something to her, as it had since our high school days of activism. The position—she was hired to write annual reports for the EPA's chief financial planner—wasn't exactly what she wanted, but she thought it would get her foot in the door, and something better would follow. But years passed, and, for all her promotions up the bureaucratic ladder, nothing did.

September 11 woke her up—and pushed her to do more meaningful work. It was time to do some real environmental protection, not just paperwork for government contracts. She got the word out among EPA friends and learned that the Office of Ground Water and Drinking Water had some new positions for engineers. *Protecting drinking water*—that sounded like important work. In 2002 she began a new job in that office while taking classes part-time at the Northern Virginia campus of Virginia Tech, working toward a master's degree in environmental engineering.

But the job turned out mostly to mean writing more regulations for the Safe Drinking Water Act that would take years to propose, finalize, and implement. It was a time of seismic shifts in water treatment, when cities across the country were switching disinfectants, not only to combat a broader range of pathogens, but also to limit the effects of chlorine and its carcinogenic by-products, like TTHMs.

A public health triumph, drinking water treatment has saved countless lives from persistent menaces like *E. coli* and halted deadly epidemics of cholera. But now there were new threats—noroviruses and something called cryptosporidium, or "crypto." Cryptosporidium, a microscopic parasite that wreaks havoc on the digestive and respiratory tracts of human beings, is not effectively removed from contaminated water by chlorine, so the EPA decided it needed to create better drinking water treatment rules.

Many cities turned from chlorination to chloramination. As a disinfectant, chloramine lasts longer than chlorine, is effective against crypto, and doesn't create as many toxic by-products. By 1998, in order to comply with the new EPA regulations, an estimated 68 million Americans were drinking water disinfected with chloramine, and several major cities, including Washington, D.C., were using it.

But there were problems—as there are with any disinfectant, depending on the amounts used, the mix of other disinfectants, and composition and quality of the source water.

"Water treatment is an art," Elin wrote in a text.

ME: I'm starting to get that.

ELIN: It's complicated.

ME: I get that too.

ELIN: Every time they tweak the rules, some new thing pops up.

The problem: chloramines made the water much more corrosive, an unforeseen consequence that didn't come to light for several years. By that time, the corrosion had leached toxic levels of lead for years.

THE D.C. WATER STORY broke in an alternative newsweekly, the *Washington City Paper,* on October 18, 2002, through the story of a resident of American University Park whose water tested very high for lead—six to eighteen times the EPA action level. The EPA scientist overseeing D.C.'s water at the time suggested that drought conditions could have raised the alkaline levels of the Potomac River, which would make the water temporarily more corrosive, and caused the service lines to leach lead into the water. It was thought that there were about thirty-five thousand lead service lines serviced by the city's water utility company, D.C. Water and Sewer Authority (WASA). A plan was implemented to replace 7 percent of those lines each year until the lead levels dropped.

That was the official plan anyway. Nobody questioned it until the following year, when WASA hired Marc Edwards, a young professor of civil engineering at Virginia Tech, to look into the inexplicable increase in pinhole leaks in copper water pipes in the D.C. area. It was even happening in new copper pipes, which were supposed to last fifty years. Edwards, an expert in copper corrosion with a degree in biophysics, began researching the rise in these strange leaks.

Out of curiosity, Edwards decided to do a test for lead, suspecting that a chemical change in the water could be causing deterioration of the D.C. pipes. As he would later recount at congressional oversight hearings, the level of lead on his first test exceeded his meter's range. He had to dilute the water and test again. When he did, even the diluted water tested at a minimum of 1,250 ppb of lead, a level that Edwards later said "would literally have to be classified as a hazardous waste."

Edwards had stumbled by accident onto the unforeseen consequences of the new EPA regulations, the very ones that Elin's office had been working on. The switch to chloramine, now required by Safe Drinking Water Act regulations, caused increased corrosiveness in the water. Old service lines had been coated with decades of accumulated mineral deposits that formed a protective coating inside the pipes, but that protective layer was now being stripped by the corrosive water. And lead was leaching out.

A cover-up was already under way. When Edwards brought his findings and concerns to WASA, he was told his funding and future access to monitoring data would be cut off. At the EPA, his subcontract was discontinued. Other engineers who drew attention to the toxic lead levels and recommended the utility and the EPA take much more aggressive action, like WASA water-quality manager Seema Bhat, were fired. Elin attended the first few meetings in 2003 to address the elevated lead levels, and when she began asking a lot of questions, she was mysteriously removed from the project and promoted to another division within the agency.

Weeks, months, and finally years passed. Lead flowed freely and in heavy amounts in all four quadrants of the District, from Georgetown and Spring Valley to the farthest reaches of Georgia Avenue and Anacostia. It affected infants, children, and adults; rich, poor, and gentrified; working, middle, and upper class; white and black.

Both WASA and the EPA knew, and did nothing.

In January 2004 *The Washington Post* published its first story, reporting on the elevated water-lead levels uncovered by Edwards. At this point, not even the D.C. mayor and council members had been told. In response, WASA issued a health advisory to pregnant women and young children in homes with lead service lines—but high water-lead levels were later found in homes with copper service lines too.

As Edwards testified to Congress in March 2004, the corrosive water was leaching lead not just from the old lead service lines outside homes but from brass fixtures inside homes as well. It turned out that brass fixtures, even when marketed as "lead-free" by the manufacturer, could still contain an average of 8 percent lead. That wouldn't change until a new regulation took effect in 2014.

As the evidence became public and overwhelming, EPA and WASA administrators hunkered down, more interested in self-preservation than in public health. These agencies were joined by the CDC, which cooked up its own corrupt study and issued a statement in March 2004 that there was no evidence any child had been harmed by the elevated lead amounts.

To make matters worse, D.C. didn't want to pay to replace entire lead pipes, so it recommended the "partial" replacement of lead service lines. But in a partial replacement, the disruption of lead scale in pipes caused even more elevated lead levels in the water.

Marc Edwards, now stripped of funding and discredited by the utility and two government agencies, continued his research independently by working with homeowners directly and mortgaging his house to pay for it. Ultimately, his work and science were irrefutable. In 2008 he was awarded a MacArthur "Genius Grant" for his research. The following year he published a report on the D.C. crisis in *Environmental Science and Technology*, which tied the elevated water-lead levels to raised blood-lead levels in children. It garnered every piece of scientific acclaim under the sun.

What he discovered confirmed his worst fears. As many as 42,000 children in D.C. had been in the womb or under two years of age when they were exposed to harmful amounts of lead in drinking water from 2000 to 2004. The effects were irreversible—and almost impossible to prove.

No proof, no blame. As Elin had told me, the wrongdoers in the

D.C. crisis escaped conviction or consequences. Many were even promoted. There were investigations and lawsuits but none were conclusive. Proving that lead had had harmful and lasting effects on specific children in D.C. was harder than proving that tobacco causes cancer. Without such a smoking gun, the wrongdoing could go unchecked. And it did. Nobody went to jail. Nobody even lost their job.

Meanwhile, the public health community continued to believe the CDC mantra: that lead in the water could not harm kids. The CDC carried a lot of credibility. Everybody knew what eating lead paint chips did to kids, but without ironclad studies to disprove the CDC's assurances about lead in water, the misperception continued. And the CDC, now invested in a cover-up, did nothing to reverse its claims until it had to. In 2010 a congressional investigation found that the federal agency had made "scientifically indefensible" claims that the lead levels in D.C. were not harmful—and had knowingly used flawed data. The CDC issued a statement the next day acknowledging errors, and made a correction.

And what about the children? What about the ones who received the full impact of the lead crisis—younger children, whose brains were the most vulnerable? How are they doing? Based on what we know about how lead affects developing brains, we can make some assumptions. Some may have experienced inexplicable developmental delays, behavioral problems, low test scores, and blunted potential. Some today may have neurological effects, cognitive impairments, and maybe even higher aggression and tendencies toward violence. Adults across the District may be experiencing an elevated risk for memory loss, early dementia, hypertension, gout, and retinal hemorrhages.

But these things were never investigated.

AROUND TWO IN THE MORNING, I put down my iPad, trying to process it all. I couldn't remember when I'd stayed up so late. But I was confused and reeling from what I'd read. Why wasn't the D.C.

crisis better known? It was a public health tragedy—the kind that government regulations are meant to protect us from—that occurred right under the noses of our most powerful institutions. Why hadn't it changed everything?

The next day, in a serendipitous twist of fate, I had a meeting scheduled with the Genesee County Health Department, initiated by none other than the health department employee who was in charge of lead in the county. He had been awarded a grant—or the county had—to provide cleaning supplies to any family with a young child who was tested and shown to have elevated lead levels in their blood. They had designed prescription pads with information about lead abatement and cleaning supplies that we were supposed to distribute to affected families. We were meeting to figure out how to implement the program and coordinate work between the health department and the clinic.

Wild luck, right? The county expert on lead, even if he did restaurant inspections half the time, had to know something about children's lead levels. Whether the source was paint or drinking water, it shouldn't make a difference, right? And he'd have advice for me. For all I knew, the entire problem could be solved in a day. One meeting, one word with the right official—sometimes that's all it takes. Maybe this guy could push the right button and make all the lead disappear.

I picked up my tablet again and sent an email out to the three residents from my Community Pediatrics rotation who would join me at the county health department meeting later that morning. It is my habit to take residents with me to county or city meetings, since advocacy is best seen up close and in action. I sent them the articles and links Elin had shared with me and asked them to please read them before the meeting.

Then I started to write an email to my resident Allison. Just a week before, we'd seen little Nakala together. I kept thinking about how I'd told Nakala's mom that the tap water was okay.

Water. It is the most essential substance on earth. Seventy-five percent of our bodies is made up of water, and an even bigger percentage of a baby's body. It helps maintain body temperature and blood volume. It removes waste from the body, lubricates joints, and protects tissues. For babies on powdered formula, water is pretty much everything they consume, at a time when their brains are developing the most.

I wanted to get a case of bottled water to Nakala's mom. Even better, we needed to issue a statement to all our patients, to all the kids. Was it too early in the morning for that? As I drifted off to sleep, I composed memos and texts and emails in my mind, a relentless parade of words and warnings.

But I needed to stay calm. I needed to be strategic and careful. I needed to know more. The way to win this fight would be to keep a cool head.

—

CHAPTER 6

First Encounter

I KNEW AS SOON AS I WOKE UP THAT MY MOM WAS DOWN-stairs. The sweet smell of breakfast had reached the second floor. She had arrived before dawn, while we were sleeping, and was already in the kitchen, making crepes, beating up eggs, flour, butter, and milk, then pouring the batter onto the electric crepe maker. One by one, she produced the thinnest, most delicate pancakes, Nina's and Layla's favorite breakfast. My mom expressed her love in many ways, but as with most Arabic moms, love began with food.

Across the room, Elliott was asleep in the reclining chair, the ice machine still humming away. On the bed next to me was my iPad, where I'd tossed it, in a fit of disgust and fatigue, around two in the morning. I felt a twist in my stomach when I looked at it. I showered, dressed quickly, pulled my hair back into a bun, and went downstairs, eager to get to work. My meeting with the guy from the health department was first thing, at 8:30 A.M. That was all I could think about.

My mom greeted me in Arabic, "*Sabah al-kheir, habebtee.*"

Her voice was, as always, delicate and bright. We hugged, as always, even if we have just seen each other an hour before, even if I am just coming back from the grocery store. When we talk on the phone, which is often two to three times a day, it feels like we hug first. She is shorter than I am, tiny, barely five feet tall, topped by awesome choppy white hair that beautifully contrasts with her tan brown skin.

"*Sabah al-noor*, Mama."

She was drinking Iraqi *chai,* a superstrong and supersweet concoction that's brewed with loose black tea leaves and sometimes cardamom pods, then doused with an insane amount of sugar, enough to form a crystalline layer at the bottom of the clear glass cup. Over the years, my parents cut the sugar, but they never cut back on drinking *chai.* They brewed it every morning in a ceramic pot insulated with a handmade tea cozy and put on an electric warmer, where it sat for the day. As with most Iraqis, *chai* was an obsession with them, consumed at all hours, even on the most blistering summer days. As kids, Mark and I drank *chai wa haleeb,* a warm and creamy brew with lots of milk, the taste of which takes me immediately back to childhood. It wasn't until Mark got to law school, and I got to medical school, that we moved on to the higher amounts of caffeine that modern life seems to require. As comforting as the smell was, today was not a *chai* day.

As I sleepily made myself a strong dark coffee, my mom lifted a steaming crepe from her crepe maker and began making another. Even in retirement, she was unstoppable in many ways—still knitting a new sweater every month, reading a couple of novels a week in Arabic or English, celebrating each holiday and birthday with special cakes and delicious dishes that she made better and faster every year. In our first years in America, when Mark and I were little and she was so lonely and homesick, she would make sudden appearances in my school classroom, her English still rough and uncertain, carrying large platters of sweets and baklava that dwarfed her body.

Once we left the Upper Peninsula, after my dad finished his postdoc and was hired by GM, she changed. She was quicker to laugh. Her warmth and positive energy made it easy for her to make friends. My parents rented a tiny apartment in Royal Oak, a suburb of Detroit, and began building a new life, joining the vast Arab community all over Metro Detroit. In a year they had enough money, some loaned by a cousin—because they couldn't get a real mortgage—to buy a small one-story brick house from a Palestinian friend on Mark Orr Road, where the cookie-cutter houses were set close to one another. That's where we lived for the next fifteen years, where Mark and I spent our childhoods. On summer days we rode our bikes to Memorial Park and to Ray's Ice Cream, a shop right out of the 1950s, where the line snaked out the door and down the sidewalk.

Education was the religion of our family, embraced as a way to a better life but also to a richer, more intellectually alive existence. Doing schoolwork and getting good grades were expected by my Arab tiger mom. But in addition to what we learned in school, my parents urged us to read and learn independently about geography, history, literature, international affairs, and current events.

WE PAID A PRICE for our new American lives, in ways that were often unexpected and painful. In Royal Oak, there were only a handful of other minority kids in our schools. Mark and I, while growing up there, rarely talked or obsessed about being called "camel jockeys" and other ethnic slurs. Though these incidents were infrequent, they did seem to coincide with U.S. military actions against Arab countries, usually Iraq, that kids were hearing about in the news.

I'm sure my parents had no idea, before they settled in Royal Oak, about the working-class suburb's dark history as a base of operations for Monsignor Charles Coughlin, usually shortened to Father Coughlin, and his angry, hate-spewing, anti-Semitic radio program, broadcast on Sunday afternoons throughout the 1930s—first on WJR, then nationally on CBS. At the time, with tens of millions of weekly listeners, he was one of the most influential voices in America.

Coughlin promoted the racism and nationalism of Hitler and Mussolini, offering weekly installments of *The Protocols of the Elders of Zion*, the forged document that became the *ur*-text of anti-Semites. His newspaper, ironically named *Social Justice*, evangelized for America First–ism, a rudderless economic populism mixed with isolationism. After the outbreak of World War II, FDR's administration declared his hate-filled broadcast "enemy propaganda" and created new broadcasting restrictions specifically to force him off the air.

You would think my history classes in Royal Oak might have bothered to teach this. They didn't. Instead, the working-class vanilla suburb of Royal Oak that Mark and I knew was universally called "Royal Joke." It was a mecca for angsty punk teenagers with red Mohawks who hung out downtown near the Noir Leather fetish store, where I bought my first pair of Doc Martens. It was where indie coffee shops and grunge first arrived in Metro Detroit.

Years later I learned that the pretty Catholic church on Woodward Avenue, National Shrine of the Little Flower Basilica, had served as Father Coughlin's headquarters. And I was stunned to discover that the large post office in downtown Royal Oak—where our family mail was sorted, and where I dealt with bulk mailings for the political campaigns of John Freeman and my high school friend Dave Woodward—had been built to process the crazy amounts of fan letters Coughlin received. According to historian Alan Brinkley, in 1934 Coughlin received more than ten thousand pieces of mail a day.

We don't think enough about what lies beneath the veneer of the places where we grew up, as if childhood innocence lingers inside us, filtering out anything too complicated or too dark to consider. We step over complex systems every day, walking through history and pretending the darkness isn't there. But the older I get, the more I want to really understand the world I'm in and how it came to be. I learned that from my parents—to dig deeper and not be afraid of what I might find.

——

AIRMAIL LETTERS ARRIVED FROM overseas for my mom, their thin, see-through envelopes striped with blue and red slashes, the words *par avion* on the front. Inside, the blue-lined, extremely narrow paper was crammed with lots of writing in a little space. Sometimes Mark and I came home from school and noticed one of these letters open on the counter and were immediately filled with dread.

As excited as my mom was to receive them, the letters often contained bad news, things happening six thousand miles away that she couldn't do anything about. A letter might tell her that her brother or brother-in-law was being deployed to the front during the Iran-Iraq war, a senseless and drawn-out conflict that killed more than one million people between 1980 and 1988. Or she'd learn that a favorite uncle was dying. I would arrive home thinking about homework or a weekend party with friends from SEA, and the next thing I knew, I was hearing about cousins dealing with air raids and food shortages, or about an old family friend who had disappeared. My mom would read and reread these letters, sometimes for hours, crying. There was still homemade food in the kitchen, ready for us. But her mind had traveled elsewhere.

Things were easier in the spring, when she was consumed by our garden, planting roses in the front and growing a large vegetable garden in the back—eggplant, peppers, *tarouza*, or Iraqi cucumbers, and so many tomatoes that by the end of August, she harvested enough to can for the winter, so our *maraka* always had the best flavor. Along the side of our house, near our bedroom windows, she planted honeysuckle, white and yellow buds that blossomed in the spring and filled our rooms with the smells she remembered from home. The number of plants and containers and smells in our garden seemed to multiply each year, forming a lush jungle that grew all summer until my mom moved the plants back inside for the long Michigan winter, when our dining room became a greenhouse. The plants grew as high as the ceiling and drove my father crazy. My mother slowly re-created elements of her distant home in ours.

But the legacy she aimed to create wasn't just in the physical space we lived in.

MY MOM AND ME IN FRONT OF OUR CHILDHOOD HOME IN ROYAL OAK

After five or six years at his job, my dad was promoted to the highest research position, "technical fellow," at the GM Tech Center in Warren, a space-age building designed by architect Eero Saarinen. GM's informal 1960s slogan—"what's good for General Motors is good for America"—was becoming a little outdated, but not in our house. What was good for GM was certainly good for our family. The company was still a vibrant innovation machine, much like Bell Labs, IBM, and Kodak used to be, and gave my dad the freedom that a creative scientist and inventor can usually only dream about, along with financial stability, a pension, and health benefits. In return, he designed and developed new metal tools and parts, alloys and applications—acquiring forty-three patents in his name. In time, though, he would watch the greatest international manufacturing company in the world slowly fall apart.

Eventually my mom returned to college to validate her chemistry degree from Baghdad University, getting a master's in chemistry and a teaching certificate at the same time. She discovered that ESL (English as a Second Language) teachers were in demand all over the state, and before long she had a job as a "parapro," or paraprofessional, similar to a teacher's aide, at our high school. (Mark and I joked that she worked there to spy on us.)

Once certified, she thrived as an ESL teacher, tapping into her own experiences as a reluctant immigrant. She worked for years helping young students, mostly from Japan and the Middle East, including refugees from war-torn Iraq. Along with teaching them English and getting them through their textbooks and coursework, she helped them acclimate to their new lives, explaining American customs like school musicals, homecoming dances, and the prom.

As it had for so many immigrants over the centuries, the promise of America worked for my family. We'd left a country that was broken, unsafe, unpredictable, and oppressing its own people for a country that allowed us to thrive. My parents didn't have much when they arrived in the United States, but they were able to use their educations to find good-paying jobs, buy a house in a safe neighborhood, and educate Mark and me at Michigan's excellent public schools and universities. The American Dream—buoyed, backed, and underwritten by the choices of the American people, expressed through their democratically elected government— worked for us in so many ways that it no longer works for my kids in Flint—and maybe was never meant to.

MY MOM NEEDED A LOT of convincing before she retired. She finally took a buyout from the school district, which was a big relief to me. She did it just in time, right before the state legislature began limiting pensions and benefits for teachers who had retired during the Great Recession, which started earlier, hurt more, and lasted longer in Michigan than in the rest of the country.

I needed her. We all did. Elliott was now fully immersed in the world of school health while also overseeing a mobile medical program that treated kids in Detroit. And the jobs I wanted were intense and required a significant time commitment. From our point of view, the more time Nina and Layla spend with my mom, the better. She speaks to them in Arabic. She teaches them to cook and knit, when they are interested. She brings the culture of old Iraq and our family stories into their lives. She is even willing to deal with the logistical hassles of carpools and field trips and school volunteering. After I took the job at Hurley, Elliott and I bought a house just ten minutes away from my parents—halfway between Flint, where I worked, and Detroit, where Elliott did. This was, I recognized, another privilege our family enjoyed.

Bebe. That's what the girls call her. The love between them is immeasurable and precious. She makes every sleepover an adventure—with Audrey Hepburn movie festivals, card games, baking adventures, and craft projects. She doesn't just look after Nina and Layla. She helps us raise them. And she looks after Elliott and me too.

As for my dad, or Jidu, as the girls call him, he never really stopped working. The layoffs and reorganizing that hit the assembly lines of Flint so hard in the 1980s slowly worked their way to Warren and my dad's fantastic space-age office. The economic crisis in 2008 began waves of layoffs. In one day, hundreds of engineers and scientists were fired—instructed to empty their desks and leave the building where they'd worked their entire lives.

The layoffs sometimes seemed random to my dad, and sometimes unfair. My dad would come home with sad stories of having to say goodbye to people he'd worked alongside for decades. He believed in GM, perhaps too much. As the stock price went down, he bought more GM stock. With the company bankruptcy in 2008, the "old GM" stock literally disappeared, and so did quite a bit of his 401(k). Yet he continued to have faith in the company that had put food on our table.

When he survived the layoffs in 2008 and 2009, we rejoiced. After he survived more cuts in 2010 and 2011, it became a family joke that he was on some sort of "safe list." What did Daddy have on GM? Why was he still there? Finally, just a few months back, after thirty-one years, he retired and began a new life with new kinds of work, from obsessively researching our ancestral history to indulging Nina and Layla as well as Mark's two boys, Theodore and Zachary, with expressions of love and affection that had been largely absent during my own childhood.

I chalked it up to the softening that comes with age and time, the ability of love to heal, and the mysteries of being a grandparent.

His activism hadn't died or faded. It morphed. His view shifted. He embraced what was in front of him—his love of his family, his realized dreams for Mark and me, his hopes for our kids—rather than holding on to what was behind him or across the sea. His transformation grounded me, nourished me, and gave me hope. Time and love had brought out the best in my dad, buffering the trauma of his past. Resilience, I knew firsthand, could be learned.

"How was the barbecue?" Bebe asked me. Over the years, she and my dad had managed to charm and semi-adopt many of my friends, but Elin was a particular favorite. My mom had always been impressed by how academically and musically accomplished she was. When my mom visited D.C., she always made food for Mark and his family—and for Elin too. Later, when Elin moved back to Michigan, my mom gave her cooking lessons. The fact that Elin's dad and mine had both worked at the GM Tech Center for years created even more shared history.

"Oh, it was fun," I said. "It was nice to be together."

"What did you serve?" she asked.

I described the meal briefly, knowing that whatever I said, my mom wouldn't think it was enough food. She hasn't quite given up on my domestic skills, but she was already turning her attention to Nina and Layla, hoping they might be more receptive to her cooking lessons than I was.

"Everybody seems good," I said. "Elin says hi."

I was careful to say nothing about the Flint water. I have always kept pretty quiet about my work, wanting to savor my family time, focus on the girls, and maintain boundaries. My parents and Elliott are used to it. My dad was the same way about his job at GM. But that morning, my sense of privacy, or boundaries, whatever you call it, was heightened. My mom is a chronic worrier. She worried that I might speed on my way to work or use my cellphone and become distracted while driving. She worried that I worked too hard, or that I didn't have enough fun or days of vacation. She never sheltered Mark and me from the bad news coming from Baghdad when we were little, yet in many ways she was the prototypically overprotective immigrant mom. She enforced early curfews and banned slumber parties, dating, and watching violent movies. A few broken curfews aside, I never really gave her cause to worry. And when I did give her cause, I felt guilty.

I got better at protecting Bebe once I was an adult. She is anxious about the unknowns, the hundred things that can consume a woman who has lost everything of her youth and has decided she is not going to risk losing anything more.

LAYLA SMELLED BREAKFAST AND rushed downstairs, dressed, alert, and ready for another day of Skull Island.

"Layla *oumree!*" my mom called out, a term of endearment that means "Layla, my life!" and wrapped her arms around my youngest daughter before handing her a plate of crepes with gooey Nutella on top. Several minutes later, after multiple calls of *Yalla!*—"It's time to get up!"—Nina finally appeared, groggy and quiet. She was slow to get going in the morning. She hugged and kissed my mom, then sat at the counter with her plate of crepes, piled high with strawberries and cream.

"Did you get enough sleep?" Bebe asked me.

"No, not really," I said distractedly, immediately regretting having said it.

Luckily, Elliott popped into the kitchen, his arm in the sling and his hair disheveled. He was the perfect distraction. I grabbed my go-cup of coffee, kissed everybody goodbye, and called out, "Bye, *habebtees*! I love you. Have fun at camp!" And I headed straight to the garage.

It was another chilly morning, and damp. On I-75, the same road where I had been in that car accident all those years ago, I drove to Flint, my usual commute, feeling obsessed and stressed. My mind had focused narrowly on one thing and one thing only.

FIVE OF US WERE sitting around the large oval conference table, down the hallway from my office at Hurley Hospital, in what used to be the old pediatric psych wing. The room felt too big with so many empty chairs and dead space. For the meeting with the guy from the Genesee County Health Department, I was joined by three residents, all of us in white coats.

The guy from the county health office was in his thirties and dressed in the business-casual style of someone who does restaurant inspections half of the time. He had recently come to Genesee from a nearby county's health department, he said, with twice as much funding and only half the population. That was annoying to hear. As soon as we started talking, he complained about how few resources he had for lead remediation in Genesee. He said his main focus was on trying to build a stronger and better public health program.

I had no argument with that goal. Funding for public health is inconsistent in Michigan, which has a weird system that just doesn't make sense. In the past, the state took on more responsibility for county and local expenses, but especially under Governor Snyder, there were major cuts in revenue sharing to local governments. Consequently, counties and municipalities relied on their own tax bases, mostly from property taxes, to maintain services.

The result was wild disparities in how effectively the health of Michigan residents was protected. When money was tight, public health budgets were often the first to be reduced. In Genesee County, for instance, where the crime rate was considerable and the sheriff wielded immense power, most of the resources went to the sheriff's department, for better offices and more staff, guns, uniforms, and squad cars. The health department got less.

It made no sense that communities with the most struggles and most poverty—and therefore the most health issues—were always allocated the least amount of money, but that was how it worked, since the property tax revenue was smaller in poor counties. A similar dynamic affected school funding in Michigan, allowing the richest school districts to spend more per student, especially on capital improvement or school buildings. Everyone in government knew about this inequity—in fact, it was the legislature's choice to set up the system like that—but the unfair situation was passively accepted with a "well, life isn't fair" shrug. And the state government's cuts to local revenue sharing made things much worse. In Flint, despite years of state oversight of the city and its services, the budget deficits had never been closed, not even by the parade of EMs.

When a city doesn't generate enough tax revenue because property taxes don't bring in enough money, the poor people who live there are punished with higher utility bills. It's very regressive thinking, asking poor people to pay a higher share of their income than other residents for basic public health protections like water or adequate plumbing. Flint had miles and miles of old pipes underground that needed repair and replacement. In 2014 the city pipes were leaking between 20 and 40 percent of their load, which meant residents and business owners had to pay for those water losses. The average annual Flint residential water bill in 2015 was $864—about $300 more than in any other city in Michigan. In fact, it was the highest in the nation.

The guy from the health department and I moved on to talk about making sure the families with children who tested for high levels of lead would have easy access to the cleaning supplies. Lead poisoning from old paint and paint dust was an ongoing concern in Flint, due to the ancient housing stock that was often in disrepair. What these houses really needed was a serious inspection and total lead abatement, but that was more money than the government wanted to spend. So instead, we offered mops and cleaning supplies.

I felt a pang of frustration but held my tongue and waited until the end of the meeting to raise the subject of the Flint water. First I brought up the recent newspaper articles about the leaked EPA memo.

"Do you know what's being done about that?" I asked.

He looked puzzled.

"You've heard about the water problems in Flint—the lead?"

He shook his head.

"Testing has been done by a drinking water expert at the EPA," I went on, "and the water shows high levels."

My residents all nodded. They had read the articles I sent them early that morning.

The health guy looked away, as if anxious to leave, then turned to me and said, "Water is not under the jurisdiction of the health department."

"What do you mean?"

"Water isn't our department. We deal with lead paint and lead dust," he said. "When it comes to water, that's another department. It's under public works."

"Excuse me, it's a public health issue," I said, suddenly becoming aware of how short my fuse was. Public works? *What was that?* "I understand the focus on lead paint and lead dust, which, don't get me wrong, is a serious problem. But lead is lead, no matter the source. How can *lead in the water* not be under the jurisdiction of the health department?"

He shifted in his chair.

"Doesn't your office collect all the blood-lead-level data for the county?" I asked.

He nodded.

"What about the lead levels of kids? Have you noticed any changes in the levels over the last year? *Has anybody noticed?*"

"I don't know if anyone has looked," he said.

I wasn't getting anywhere, I could see that. The guy had no power, and he didn't plan to shake the cage or step beyond the confines of his job description. The meeting ended pretty quickly after that.

I soothed myself with a new plan. I would fire off an email to his bosses at the county health department. I didn't really know them, but I knew their names. Surely somebody there knew something and could help.

—

CHAPTER 7

Miasma

YOU CAN BE DRAWN TO THE FIELD OF PUBLIC HEALTH AS A humanitarian, a mathematician, a statistician, or a health provider. It can be a religious calling, or it can spring from a passion for pure science. For me, except the religious business, it's a combination of all those things, along with my love for a bit of suspense. Addressing a public health crisis or curing a disease is like solving a mystery, usually with just the right mix of instinct, insight, footwork, solid data, strategy, and pure luck. The history of public health is loaded with incredible stories and puzzles, which is why so many books are written about contagions and outbreaks. Even zombie stories are metaphors for epidemics.

Before public health departments became part of government, responsibility for people's welfare fell on their families and communities. It's been only fairly recently, over the last couple of centuries, that governments became involved in protecting the health of citizens. As a field of science, public health rose up with the cities that formed around the time of the industrial revolution. As people crowded into smaller geographic spaces, disease spread faster. Wood and coal fires polluted the air. Human waste ran into the streets. Running water was very rare, sewers even rarer.

My favorite sleuth is John Snow, a nineteenth-century physician, inventor, scientist, advocate, and founding father of public health. The son of laborers who worked his way through school, Snow first made his name as an anesthesiologist, before that was even a discipline, by inventing new ways to ease the pain of patients during surgery. He was even called to Queen Victoria's bedside in childbirth. Snow published dozens of papers on a wide range of public health and medical issues, including lead poisoning. But more than anything, Snow loved systems, networks, demographics, and, most of all, epidemiology—the branch of medicine that deals with the spread and control of disease.

A century ago the biggest threat to life wasn't cancer or heart attacks—it was infectious disease. Nothing surpassed his obsession with preventing cholera.

The effects of cholera are devastating. In a healthy person, the small intestine absorbs more water than it secretes, which keeps the cells of the body hydrated. But an invasion of *Vibrio cholerae* reverses that balance. A victim experiences sudden diarrhea, vomiting, and severe dehydration—quickly evacuating all fluid from their body. In a matter of *hours* after contracting the disease, the victim often dies. In the nineteenth century, the fluids left behind contaminated the sewage systems, then the drinking water, and spread the disease.

In London's first outbreak, Snow's own records show that 4,736 lives were lost. Seventeen years later, in 1849, a second outbreak in London claimed another 14,137. When cholera struck yet again, in the summer of 1854, victims in the crowded neighborhood of Soho began dying—eventually 10,530 of them.

A combination of luck, location, hard work, and brilliant instincts led John Snow to a major scientific breakthrough. Learning from his work and experiments with various anesthetics, all inhalants (primarily ether and chloroform gas), Snow doubted the prevailing "miasma" theory that cholera was spread by breathing stagnant air. He argued that people working in all kinds of smelly places weren't getting sick. He suspected that the disease was spread by unsanitary drinking water instead. Snow conducted research, published papers, and gave presentations to the London Epidemiological Society. But few doctors in London at the time felt comfortable bucking consensus. And politics played a part.

Miasma was the established theory—and had the support of the entire medical establishment and public health community, including the persuasive reformer Edwin Chadwick, who had led a charge to clean up the foul and stinky air in London by creating a drainage system for under-house cesspools, a predecessor of sewage systems. But the expensive new system relied on an ancient drainage network underneath the city that led straight to the River Thames, the source of the water that South London used for cooking, bathing, and drinking. Chadwick's new drainage system actually left the population far more vulnerable to disease.

When the 1854 cholera outbreak occurred in Snow's own Soho neighborhood, he once again made his case, this time to the Board of Guardians of St. James's Parish. He argued that the Broad Street pump, where many people in the parish got their water, was contaminated—and was spreading the disease. As far as the eye could see, the water seemed clean, with no smell or any other sign of danger, but it could kill you if it had enough cholera in it. The board doubted Snow, but the pump handle was removed (just in case). By then, most people had fled the neighborhood or were using water from clean wells nearby. The outbreak was stayed.

But protecting his neighborhood—and his own reputation—wasn't enough for Snow. His passion for science led him to try to uncover how cholera is transmitted, to prevent more outbreaks in the future. So he set about proving his theory, teaming up with Henry Whitehead, a young clergyman from another local church who believed the miasma theory. Working parallel tracks, they conducted parallel investigations and the two men created a "ghost map" where each black square or bar (later modified to dots) represented a cholera case in London. The map demonstrated an undeniable and predictable pattern: the black squares appeared in clusters close to the Broad Street water pump. Some black squares showed cholera victims who lived farther away, but Snow was able to prove they'd consumed the water from the Broad Street pump too.

SNOW AND WHITEHEAD'S NEIGHBORHOOD MAP

If it weren't for Snow's science, stubbornness, persistence, and passion for the truth, cholera might have raged on for another decade or more, taking thousands or even millions of lives. When I think about Snow and his accomplishment, what has always grabbed me most—and impressed me—is the way he insinuated himself into the epidemic. Nobody hired or paid him—or even asked him—to solve this epidemic. But he had a crucial tool at his disposal, epidemiology, and a problem right in front of him, in his own neighborhood, and that was enough for him to get started. He didn't stop his research after the pump handle was removed. He kept going, kept pushing, kept researching to prove his theory and leave a lasting contribution. As a citizen and physician, he felt duty bound to share his work and make a difference, to prevent future epidemics and save lives.

It may have been his humble background that drove him. He never blamed the less fortunate for their predicaments; instead, he understood and studied the way that their environment—whether it was poor light or lack of running water—contributed to their condition.

Rather than standing on the sidelines, Snow got passionately involved. His work wasn't about abstract scientific discovery alone. It was about people and community. That's what science is supposed to be about—not an academic exercise for the ivory tower, or racking up publications, grants, and offers of tenure. It's about using the tools and technology available to make lives better, no matter what articles of faith obstruct the path.

Speaking science to power, Snow was a disrupter of the status quo, not for disruption's sake alone, but for people. Snow was so far ahead of his time, his work wasn't totally vindicated until years after his own premature death.

I LOVE STORIES ABOUT people who can—simply by being persistent, methodical, and dedicated—change the trajectory of a life or even an entire population and generations to come. And what I love in particular is a good, engrossing public health mystery, perhaps because I have one in my family. The person at the center of it isn't famous like John Snow, but a distant cousin of mine, a bacteriologist named Paul Shekwana. He was one of the first public health scientists from the Middle East, from present-day Iraq, to work in America. He was a "bacteriologist" back in 1904, which is—I'm pretty sure—what we would call a microbiologist, epidemiologist, or infectious disease expert today. After studying at the Royal College of Physicians and Surgeons in England—where he worked on sewage and water testing for bacteria and might have even traded water samples with John Snow—he was hired in 1904 by the department of pathology at George Washington University in D.C.

Almost immediately after he got to America, he was called to Iowa City, where a deadly outbreak of typhoid fever had struck. Shekwana was brought in to work with the Iowa State Board of Health bacteriology lab—an entire floor of the new Iowa City Medical Building was given over to his lab team. There Shekwana investigated, among other things, the tie between unpasteurized milk and typhoid. But he didn't stop there; he promoted new public health regulations in Iowa and beyond.

His most important contribution may have been an article published in the *New York Medical Society Journal* in 1906 (which was excerpted in the *Journal of the American Medical Association*), urging all doctors to wash and disinfect their hands throughout the day,

DR. PAUL
SHEKWANA,
MY DISTANT
COUSIN, 1904

particularly before and after seeing patients. It's almost impossible to imagine how much this simple practice improved patient care, prevented the spread of infection, and saved lives, lots of them. But even so, hand-washing rates in hospitals still have to be monitored and have much room for improvement. According to a recent review, as many as one million lives could be saved worldwide each year if more people washed their hands.

From typhoid fever to hand hygiene, Shekwana was a roving public health warrior throughout the Midwest. His comings and goings were regularly reported by the *Iowa City Press-Citizen* and the *Iowa City Daily Press,* as well as newspapers in other cities that he visited. He wrote articles about sanitation, food safety, and even drinking water. Amazingly, my family has a letter in which Shekwana urges residents not to use a certain well because of the "variations in quality" of the water.

Hopefully the eerie similarities between his career and mine don't include an untimely and mysterious death. In the summer of 1906, Shekwana announced he was returning to England—there is no record of why—and resigned his position at the Iowa State lab. One afternoon before he left, he went fishing and walked back home along a railway trestle outside Cedar Rapids. He was found below the trestle—having leaped to his death or been thrown there by an oncoming train. A broken rib had pierced his lungs. He died hours later. The conductor of the train that may have killed him claimed he wasn't able to stop soon enough to avoid hitting Shekwana, nor did Shekwana try to step away. An investigation was conducted. There was no proof that the train hit Shekwana, who, it was suggested, may have killed himself.

One hundred years later, the mayor of Iowa City honored him with an official proclamation for his work in public health. Was he murdered? Was it a suicide? The Paul Shekwana story has mesmerized my family for years. At the time, his friends and colleagues described him as cheerful and excited about returning to England. Perhaps he had a love interest they never knew about. Perhaps his heart was broken. Or maybe he had caused too much trouble with all his bad news about germs, the spread of infectious disease, and water quality. As with all things, there is so much more to know.

SNOW'S AND SHEKWANA'S DAY, so different from our time, is best captured by Charles Dickens, who as a novelist was also a social critic and child advocate. His eyes wide open to the dark alleys of the industrial revolution, Dickens captured the gross inequities of his time and always kept the most innocent at the center of his stories.

"In the little world in which children have their existence," he wrote in *Great Expectations* (1861), "there is nothing so finely perceived and so finely felt, as injustice. It may be only small injustice that the child can be exposed to; but the child is small, and its world is small." The feelings of children are as vulnerable as their health. And the injustices of Dickens's time played out over a lifetime. In 1842 the average life span of an upper-class "gentleman" in London was forty-five years, the average tradesman lived to be twenty-five, and an average member of the working poor died at sixteen. Among the recorded deaths of the same year, 62 percent were children under the age of five.

Urban poverty is less lethal now, but in some respects, nothing has really changed. The environments of the cities we live in—their dirt and air, their violence and hopelessness and stress, their water— can still predict how long a life we will have. What we ingest or experience or inhale will make a difference to our health—literally the number of minutes allotted us to live. The small boy in Dickens's novel may feel injustices keenly and remember them for the rest of his days, but so does his body. It carries the injustice forward with him, always.

AFTER MY MEETING WITH the powerless county health guy, I had a full day of work. It was a Thursday, which meant that I was supervising the residents who saw patients that afternoon. My phone kept pinging and buzzing throughout the morning as Elin and I exchanged texts and emails. Our night of talking had sent her into a dark tunnel, reliving her experiences in D.C. It was like she was suffering from Drinking Water Crisis PTSD.

She asked how my meeting went, and I quickly texted back my plan B: I was going to follow up with the Genesee County health officer and medical director. The pattern of mistakes was already becoming clear to me. It was like we were reenacting the error of Edwin Chadwick's mission to clean up London's air by draining all the human waste into the drinking water of the Thames. But in this case, the mistake had been made in an attempt to save not lives but dollars.

As our exchanges went on, Elin and I started to finish each other's sentences in a mash-up of guilt and fury.

ME: It's an ignorance-is-bliss system

ELIN: Too expensive to replace the lead service lines

ME: But everyone thinks their tap water is safe

ELIN: The system is supposed to work

ME: But it doesn't

LATER THAT SAME DAY, August 27, I sent my email to the county health guy and copied in a bunch of his bosses, including Mark Valacak, the county health director. I explained that even though the water might not be under the health department's jurisdiction ("however, I don't understand why it wouldn't be"), I was concerned about the potential for an increase in childhood lead poisoning from Flint's drinking water.

"This is strikingly similar to what happened in Washington, D.C.," I wrote—a crisis that had "resulted in significant childhood lead poisonings."

I urged them to collaborate with me to stop what could be another crisis. Our children were already dealing with so much, every measurable health disparity. Adding avoidable lead exposure to their burden was unconscionable. Just to make sure they paid attention, I attached a link to Curt Guyette's scary *Deadline Detroit* article.

Poisoning is poisoning. I couldn't imagine how the county health department could turn a blind eye, or even a partially blind eye, to the situation simply because of bureaucratic walls and red tape. This wasn't the kind of issue where you can shrug and pass the buck. We were literally talking about the systematic poisoning of our children.

"What about the lead levels of kids?" I asked in my email. "Have you noticed any changes?"

Then I forwarded the email to Elin, who responded with more news from the water front.

Marc Edwards, the corrosion expert from Virginia Tech and warrior scientist of the D.C. crisis, had come to Flint. He had been invited by both LeeAnne Walters and Miguel Del Toral and was collaborating with homeowners on something he called "citizen testing." Just that day, he had posted new numbers on his Flint Water Study website. Scrolling down the numbers, I was upset to see that the citizen testing had found very high levels of lead in the water.

One sample, collected *after* forty-five seconds of flushing, exceeded 1,000 ppb—sixty-five times the federal action level of 15 ppb. The numbers were frightening.

Almost immediately, the spokesperson for MDEQ, Brad Wurfel, a confident, smooth-talking, square-faced corporate type (who happened to be married to the governor's spokesperson, which somehow gave him more credibility), began pushing back very hard. "Flint drinking water meets state and federal safe drinking water standards," Wurfel said.

This didn't bode well. Marc Edwards was a world expert on pipe corrosion and recipient of a "genius grant" for his visionary work and doggedness. As a scientist, he had been willing to stake his own retirement money to save kids from lead poisoning when the D.C. utility, the D.C. government, the EPA, and the CDC all denied there was a problem. Was MDEQ really going to totally dismiss his findings?

Wow. It took my breath away.

Brad Wurfel was just a PR guy, a mouthpiece, not the brains behind these audaciously wrongheaded remarks. I wondered about the people he worked for. *How can anybody who knows about lead not be concerned?*

"WHAT'S GOING ON AT MDEQ?" I wrote to Elin.

It seemed like lunacy to deny water test results like these. But Elin predicted more problems to come. Based on her experience in D.C., she felt things would get much worse before they could get better—or drag on for years and never reach clear resolution.

In my mind, it wasn't a coincidence that D.C. and Flint are both places, in different ways, that lack adequate political representation— places where democracy is far from complete. Flint had been taken over by the governor's emergency manager, but at least the residents could still vote, in Senate and House elections, for politicians who—if we could get them to pay attention—would represent them. D.C. was a different story. Even though D.C. is more populous than half the states in the union, it has no representation in Congress. The people of D.C. cannot fight battles the way people elsewhere can.

In Flint, with an unelected emergency manager in charge, the citizens experienced a similar disconnect and powerlessness. Layers of accountability and responsibility had been stripped away.

Politics is about how we treat one another, how we sustain and share our common spaces and our environment. When people are excluded from politics, they have no say in the common space, no sharing of common resources. People may think of this as benign neglect, but it isn't benign. It is malignant—and intentional.

Elin had another concern: Marc Edwards. The fact that he was sampling water in Flint meant something serious was going on, but his involvement could also make things more complicated. He had a reputation in the water world for making scenes and grabbing headlines. "He sometimes uses inflammatory methods to get attention," Elin said. "People mock him for that, but I think they are afraid of him too."

I noted her alarm, but felt like we had more to be afraid of than an eccentric water genius. We were alone in Flint, left behind, and maybe even targeted. We needed every ally we could get.

I AM ALWAYS HAPPIEST on Thursday afternoons. Finally free of my crazy meeting schedule, my calendar grid, and my paperwork, I can do what I love more than anything: see kids.

The end of summer is always busy in pediatrics. We're flooded with back-to-school physicals. That afternoon the pediatric residents were bouncing from one exam room to the next, trying to stay on schedule while still getting used to the routine of our new clinic.

Allison, my resident, saw the first patient, Brandon, an active eight-year-old white boy, and shared his case with me. Brandon's mom, a thin young woman with short hair and arm tattoos, was concerned that he couldn't sit still. She said her son's school, where he'd been going for three years, had been shut down last spring—a consequence of starved budgets and population loss. In a couple of weeks, Brandon would be starting a new school, an event that is definitely stressful for any kid. I tried to soothe his new-school nerves with a few comments about second grade and the stuff he was going to be learning.

"It's going to be fun! You'll meet new people, make new friends, and I'm sure a lot of your old classmates will be going there too."

Brandon looked at me skeptically. He couldn't hide his feelings, another thing I love about kids: it was clear he was anxious. His mom continued to describe his difficulties. His summer school teacher said he was a "space cadet." Even at home, he was fid-

gety, had trouble focusing and paying attention. Allison and I dug
into possible explanations: new-school nerves, summer boredom, a
hearing issue? I looked in his chart. Last year, his school physical
made no mention of hyperactivity. I found myself wondering about
the water.

"Do you live in Flint?" I asked him.

"Yes."

"Have you been drinking the tap water?"

"Yes."

Ughhh. We always see a lot of kids with ADHD, but lead exposure can increase its likelihood. Maybe Brandon was going to have
ADHD no matter what—it has so many causes. We gave Brandon's
mom the questionnaire to complete for ADHD screening and
asked her to bring Brandon back in two weeks. Allison ordered a
blood-lead test. Before they left, I asked Allison to go back into
their exam room and recommend bottled water to Brandon's mom.
Allison shot me a perplexed look but followed instructions.

With another resident, I saw Chanel, a twelve-year-old white
girl with a plump and slightly flushed face. Four years ago she was
diagnosed with obesity and pre-diabetes. In the last year, her mom
had died young of a heart attack—after years of being overweight
with high blood pressure—and Chanel's effort to lose weight became much more serious. Besides a back-to-school exam, she was
in the clinic for a weight check.

After the medical assistant weighed Chanel, I looked over her
records on the computer and saw she had lost ten pounds in the last
six months.

"Way to go, Chanel!" I called out, and raised my hand for a fist
bump.

She beamed, a bright sunrise of a smile breaking on her face.

"How'd you do that?" I asked. "So awesome! Did you cut back on
sugary drinks and pop?" We had talked about that at her last visit—
most of her caloric intake had been from sugar in drinks.

Chanel nodded, feeling proud. And then her dad chimed in that
they were both drinking more water.

More water.

Of course they were drinking more water.

It was something we recommended to all our patients. Flint, like many communities, was in the midst of a childhood obesity epidemic. Soda and juices don't help, which is why there are national campaigns encouraging more water.

Holding my breath, I asked, "Are you drinking Flint tap water?"

"Yes. Eight glasses a day!"

My heart sank. When I mentioned there were concerns about the water, Chanel's dad said, "But they say everything's okay."

"Yes"—I nodded—"but to be extra safe, you should switch to bottled water." Then, trying to pivot to a positive note, I applauded Chanel for her hard work and gave her another fist bump. I ordered another blood-lead level test.

A few exam rooms down, Allison and I saw Jasmine, a grumpy fifteen-month-old black toddler who had pretty serious eczema. Her mom said it got worse after bathing. That didn't necessarily mean anything—almost all rashes get worse after bathing. And we have so many kids with eczema. The atopic triad of eczema, allergies, and asthma runs in families, is more prevalent in inner cities— and is worsened by a variety of environmental conditions.

When I asked where they lived and if they were on Flint water, Jasmine's mom said they were. Sometimes, she said, the water smelled like a bottle of bleach. I prescribed hydrocortisone cream for the rough spots to calm down the inflammation, lots of moisturizing ointment to help it heal, oral medicine to limit her itching— and no more baths in Flint water.

"How am I supposed to bathe her?" Jasmine's mom asked.

"What about using bottled water?" I offered.

Jasmine's mom just stared at me, until it became almost uncomfortable. "You want me to *bathe her in bottled water?*"

She had a point, and I scrambled to think of better advice, something—anything. "Is there someone you know who lives outside of Flint where you can give Jasmine a bath?"

Her mom shook her head. "We don't have a car," she said.

I was running out of ideas. Bathing a child is not supposed to be this complicated. And it was getting hard to stay calm.

Without taking a moment to breathe, I next went to see Nevaeh, a child with one of my favorite names—it's *heaven* backward. She was a three-day-old black newborn who was coming to the clinic for her first checkup. Immediately I saw that I had a chance to get on my soapbox about breastfeeding—and gave it with added urgency. But that was a lost cause. Nevaeh's mom had made up her mind and was already giving her baby formula. The hospital had sent her home with a short supply of premixed formula. Knowing that she'd likely switch to the powdered version soon, I tracked down an extra case of the ready-to-feed formula and sent it home with her. Then I said, "And when you are finished with this, be sure to mix her powdered formula with bottled water."

Usually my afternoons in the clinic are a tonic, a chance to forget my own woes and worries and hang out with Flint kids—to provoke smiles and laughs, to chart the kids' growth, and to make their parents, grandparents, and caregivers less anxious. But that day, as the afternoon wore on, my frustration continued to build. Patient after patient seemed to be dealing with some kind of water-related issue.

In the teaching space, where the doctors type up their notes and talk to their supervisors, I asked everyone I could find—medical students, residents, and other supervising physicians who were in clinic more than I was—about the water. Were they seeing kids whose blood-lead screenings were coming back elevated?

They were. One physician's assistant said she'd gotten back a level of 7 µg/dl just the week before, from a one-year-old boy. Follow-up interviews with the family didn't produce any answers. They couldn't figure out the source of the lead.

"Actually, siblings we just saw two days ago had levels of 14 and 22 µg/dl," a resident told me. "We are just about to call the family to bring them back in."

"Keep me posted," I told her. "Be sure to ask if they're drinking Flint water."

I kept thinking about the difference between individual health and population health—treating one patient versus treating many. A doctor might see an individual child with an elevated blood-lead level, but it would take a study of many patients—a population—to figure out what was happening to *all* the kids. If one doctor alone could see all the Flint kids, maybe that doctor could start to make helpful connections. But with so many doctors in our clinic, and throughout the city at other clinics, that connection couldn't happen. Each doctor might see a few higher-than-usual lead results, but they wouldn't be able to see it as an epidemic on their own. This is why training in public health is so critical for all physicians. We need to be able to step back from the individual patient and look at the bigger picture.

What the eyes don't see. That is precisely why public health surveillance programs are crucial. They regularly monitor population-wide trends that individual doctors can't detect on their own—whether it is the flu, HIV, cancer, or blood-lead levels. This is what government public health people are charged to do. It is an invaluable way of discovering paradigms. It's Epidemiology 101. John Snow taught us this.

But even when lead exposure is demonstrated across a population, it is almost impossible to *prove* causation. Did lead in the water cause Brandon's ADHD? We will never know for sure. Did the water cause Jasmine's rash? Maybe. Exposure to environmental toxins usually doesn't come with glaring symptoms, like purple spots or even a rash. The symptoms are things like learning disabilities that have a time lag. Sometimes they don't show themselves for years or even decades. For a pediatrician on the front lines, often the most you can hope for is establishing a correlation.

The more I thought about it, the angrier I got.

Before leaving the clinic, I went digging online again and discovered a couple of things about the water switch in Flint. First, in October 2014, just six months after the switch, General Motors stopped using the water at its engine plant. The company got a waiver to go back to the Lake Huron water as its source. "You don't want the higher chloride water (to result in) corrosion," the GM spokesperson said. "We noticed it some time ago."

If the water was corroding metal engine parts, what was it doing to the ancient lead pipes under the city? This happened almost a year ago, but mysteriously no alarms bells were ringing. I texted Elin with the link to the story.

ME: See this?

ELIN: Didn't know—wow.

ME: Those bastards

ELIN: You never swear

ME: Maybe I do now

ELIN: How was this allowed?

ME: GM knew the water was bad. GM screws Flint again.

My second discovery was more idiocy from MDEQ and their spokesperson, Wurfel. Just a month before, when Curt Guyette's stories for the ACLU about the leaked EPA memo broke, Wurfel had responded with this statement: "Let me start here. Anyone who is concerned about lead in the drinking water in Flint can *relax.*"

The man was a menace. If I could remove his microphone, the way John Snow got the Broad Street pump handle removed, I thought of all the people I could help.

Relax?

Does anyone relax when they are told to relax?

Has that ever worked?

Someone should do a study.

CHAPTER 8

No Response

THE LAST DAYS OF AUGUST ARE SLEEPY FOR MOST PEOPLE, the final breath of summer before school starts and regular life returns. But there was no settling down for me.

There were some complications at home that were going to make those last days more difficult for me. Elliott was not better. He was still in pain, despite spending the last five days in a new Dynasplint that was supposed to help but didn't. A follow-up appointment with his orthopedist gave him worse news: he might need a second surgery.

Both my parents were traveling. My dad was in Yantai, China, working on a project meant to improve the metal quality in Chinese automotive parts—a consulting gig he had taken after retirement and loved. He was also indulging in his other passion: researching how our religious ancestors, the Nestorians, branched into China in the seventh century. He wrote excitedly to Mark and me saying that he'd made arrangements to see a Nestorian stele from A.D. 635. The carved stone was inscribed in both Chinese and Aramaic, the ancient but dying language still spoken by Chaldeans, the language Jesus spoke. Honestly, it was hard to respond enthusiastically about such an esoteric passion—it all seemed so old and so dead—but I did my best. And since my dad makes a PowerPoint about virtually anything he's up to, I wrote back quickly: "Can't wait to see your presentation."

Meanwhile my mom was in D.C., helping out Mark and his two boys for a week while my sister-in-law was working on a conservation project to improve national parks in Patagonia. Without Bebe, my childcare support system was on a narrow tightrope. Skull Island was over, and the girls had nothing to do. That alone wasn't a problem. In the life of an overscheduled, privileged child, a little do-nothing time can be great. But it was difficult for Elliott alone to fill the void left by Bebe. He wasn't supposed to drive. It was hard enough for him just to sleep through the night.

Monday came: August 31, 2015. I woke early and sat alone in the kitchen drinking coffee while everyone slept. The house was quiet. I was waiting for a response from the county health authorities. I was sure it would come. Over the weekend, there had been more water stories in the news. My mind couldn't let go. Elin's stories about D.C. were frightening, but I couldn't see something like that happening here, not to my kids in Flint.

These are responsible folks, I told myself. Why would anybody go into public health if they didn't care deeply about something this important? The weekend hadn't even been a "long" one. And wasn't toxic water more important than playing golf or mowing the lawn?

I set down my iPad and stared at the screen. The workweek had begun.

I imagined all the health folks arriving at their offices. Sitting down at their desks. Scrolling through their in-boxes. Surely I'd hear from somebody. *Any second now, they'll write me back.*

Any second now, I'd hear from them. *Any second. Any second.*

I changed the notification and alert settings of my phone so it would beep with every new email and text. All day I carried it around with me from room to room. It made the minutes pass even more slowly and painfully.

On my way to work, I had called my mom in D.C. to check in. She immediately jumped into questions about logistics for the week—who was filling in for her, who was picking up whom, when, where. She asked what I was making for dinner. "Are you ready for soccer?" she asked, unconvinced that I knew where the cleats, uniforms, and shin guards were put away. And she was right, I didn't.

My mom is amazing, but sometimes, despite being happily married, professionally successful, and intellectually stimulated, I have the feeling that I will never meet her expectations.

"I have a lot going on," I said, and quickly changed the subject. I made a note to find the cleats.

The hours dragged by.

Monday had come and gone.

Not a word.

THAT NIGHT I SLEPT FITFULLY—almost not at all. I could tell from his restless shuffling that Elliott was also uncomfortably awake. At some point around two in the morning, I looked over in the dark, and our insomniac eyes met.

"Are you awake?" he said softly.

"Yes."

"Are you going to tell me what's going on at work?"

I sighed.

"What is it?"

"It's bad, *Nunu*. Really bad. I think my kids in Flint are in danger. Nobody seems to care."

"It can't be that bad. What is it?"

"You heard about the complaints about the Flint water?"

"Yeah, sorta."

"It's real. And it's bad."

"How bad?"

"Lead-in-the-water bad. Leaching from the old pipes. The Flint River water is corrosive—and apparently no anticorrosive treatments were used."

"For how long?"

"Months. Maybe eighteen months."

Elliott was silent, the kind of silence that meant his heart was breaking. But his wheels were also turning. For the last five years, he had worked directly inside Detroit schools, practicing in mobile health clinics—vans—seeing an array of inner-city kids. So he knew the health issues that came along with poverty. And he knew how interconnected a child's environment, education, and health are—and how poverty brings innumerable toxic stresses that compound anything you're treating, whether it's asthma, allergies, diabetes, or lead exposure.

"This is all Flint kids, right?" he said finally.

"Yes, all the kids."

"So unbelievably sad."

"I know."

I rolled over in bed and felt tears coming. The beauty of being married to another pediatrician is how little we have to explain. But Elliott being Elliott, and me being me, we tend to fall quickly into brainstorming and debate about practicalities and solutions, rather than indulging our feelings. We spent the next few minutes trying to look into the future—to see what a population-wide lead exposure in Flint could mean, specifically for the kids.

Elliott wanted to know exactly how the government was dealing with it.

"That's just it. Nobody's dealing with it. Nobody's even answering my emails."

He could hear the despair in my voice. "You're the most stubborn person I know. Keep at it."

I'm definitely hardheaded, maybe persistent, and kind of competitive, but I'm not sure I'm stubborn.

Elliott knew the obstacles I was facing. Working in Detroit, he had been frustrated when he tried to make headway in treating kids with asthma. An asthmatic himself, an inhaler always in his pocket, he knew how disruptive the disease can be. Kids with asthma were missing school and falling behind. They needed doctors. They needed to be properly diagnosed—which isn't as simple as that sounds. And after diagnosis, they needed treatment: medication.

But for lots of different reasons, including transportation problems, the kids who needed a doctor the most had the least access to one. Detroit's once-great public transit system had collapsed, and roughly 25 percent of homes in the Motor City lack a vehicle. Just getting to a pharmacy to pick up medication could be impossible.

In the late-night conversations that Elliott and I had at the time, our brainstorming sessions, we always tried to be creative and look for unexpected solutions. Elliott eventually came up with an idea for how to overcome obstacles to asthma treatment—to team up with hospital pharmacists to get medication delivered directly to kids at their schools.

"You will think of a way forward. You always do."

"I hope so. I have to." I sighed.

We said nothing for a while, both of us trying to drift off to sleep.

Then he spoke again, as if he couldn't stay silent. "Remember, you're the most stubborn person I know."

"No, I'm not."

"Yes, you are."

"No, I'm not."

IT WAS TUESDAY, SEPTEMBER 1. It had been four days since I sent my email to the county health people. I got to my office early that morning. As soon as I got settled at my desk, I wrote to Elin:

FROM: Mona Hanna-Attisha
TO: Elin Betanzo
SENT: Tuesday, September 1, 2015, 8:43 A.M.

Do you have the exact date the water switch happened?

I'll try to get a report run on our patients' lead levels done before and after.

I knew the switch happened in April 2014, but I needed the exact date. I'm sure I could have googled it, but I liked keeping Elin in the loop. I liked having my old friend beside me, and I could tell she found our collaboration gratifying.

At Hurley, we routinely took blood samples and tested every child on Medicaid for lead—or at least we were supposed to. We did it only for high-risk populations—kids on Medicaid, kids in older homes, kids who had parents with lead-related hobbies.

Years ago the CDC recommended that all kids have their lead levels tested, but the public health victory that got lead out of gas and paint—and caused rates of lead exposure to go down steadily—also caused the recommendations to relax. That should never have happened. Because just as the CDC relaxed its recommendations, new research revealed that even the smallest levels of lead in a child's blood were more damaging than we ever thought possible. We should have been doing more screening, not less.

And the blood-lead screening rates, even in Flint, were low. As in Detroit, many kids had trouble getting to their regular pediatrician due to an array of poverty-induced obstacles, from inadequate transportation to complicated childcare arrangements. I have Flint patients who've never left the city limits; they've gone only as far as the unreliable and limited bus line allows.

Even so, Hurley had the screening data for children treated at our clinic, and it wouldn't be that hard to get it, thanks to our sleek electronic medical record (EMR) system. In 2011 we started using Epic, the Cadillac of EMR software, and being the early-adopter tech dork that I am—the first one I knew to buy a PalmPilot, even though I had no one to use it with—I had taken the training and fooled around with the system's cool features.

But our clinic data is part of an even larger pool of blood-lead data that includes most of the children in the county: the Michigan health department's surveillance program, the Childhood Lead Poisoning Prevention Program, collects, tracks, and reports children's lead levels. Unluckily, the only way I could get those numbers, I thought, was through the county health department.

It was time to follow up with them again. "Hope all is well," I wrote in an email that morning. "I'm not sure if you saw today's article on *MLive* regarding the extreme corrosivity of Flint Water."

And in an effort to be persistent but pleasant, I added: "I never heard back from anyone. I would love to discuss this further to see what we can do to protect and prevent lead poisoning in our kids."

But the county health department was still silent. Tuesday came and went. Wednesday came and went. No word, no response. By then, between Elliott's humming ice machine and all the conversations going on in my head, my insomnia was full-blown. I couldn't fall asleep or stay asleep. I had no appetite; I couldn't remember when I'd last eaten a full meal.

Elin's pessimism had seemed outlandish just a week before. But now she seemed spookily prescient. This made me worry even more. What if her prediction came true? *This could take two years to be resolved.* Those words haunted me. All I cared or thought about was getting somebody to pay attention. But so many days had come and gone, I suspected that it wasn't just laziness or distraction. I was starting to wonder if they were purposefully not getting back to me. What was really going on?

ON THURSDAY MORNING, SEPTEMBER 3, I wrote to Dean Sienko, a physician, a mentor, and one of the most experienced, serious public health guys I know. He had worked at the CDC years before, but he'd been a chief medical executive for the state, and a county health officer too. So he knew, better than anyone I knew, how public health worked at every level of government. I hoped he'd have some advice for me on how to get the county health department to finally respond.

Beyond all his mega-qualifications in health, Dean Sienko was a military man. He has a close-cropped buzz cut, broad shoulders, and the posture of a major general—which he actually was. In fact, his last military stint was as commanding general of the U.S. Army Public Health Center at Aberdeen Proving Ground, Maryland. Now he had a gig at Michigan State University as our first associate dean of public health. This meant you could actually refer to him as "Dean Dean"—or at least I did, which always reminded me of Major Major in *Catch-22*, one of my favorite books. To give Dean Dean an idea of where I was heading, I forwarded my unanswered emails.

Later that day I finally got a response from the Genesee County Health Department, but I wondered why they'd bothered. It was total gibberish, a foot-dragging nonanswer. They suggested that a research study could be started the following spring—in other words, after six more months of lead-poisoned water!

I tried another approach the next day, Friday, September 4— a last-minute Hail Mary pass before the long Labor Day weekend descended. Unable to get the blood-lead surveillance data from the county, I reached out to Michael Roebuck, an ER doctor and our chief medical information officer. I hoped he could help me get the blood-lead levels just from our clinic. He was a data dork, like me, and understood the power and potential of big data to improve patient outcomes at a population health level. He led the implementation of the new EMR system at Hurley and had it down cold. He was also a problem-solving type, the kind who said yes without hesitation because he believed anything was possible.

Beyond that, he and I were friends and loved to banter, often about our size and job differences. I'm barely five foot two, while Roebuck is over six feet. My patients are kids, while he takes care of adults. He liked to tease me with jokes about how "little" pediatricians liked to treat "little" patients, and about how in my office, I probably napped on a little cot, ate cheddar Goldfish, and drank from a sippy cup.

Omitting our usual jokes, I cut straight to my request. Figuring that food might be a way to get a faster response from him, I threw in an offer of lunch.

FROM: Mona Hanna-Attisha
TO: Dr. Michael Roebuck
SENT: Friday, September 04, 2015, 12:57 P.M.
SUBJECT: Fwd: report

I need an Epic report regarding lead levels. I'll take you to lunch. ;)

But even Roebuck didn't write back. That was a shock. Techies are always connected to devices and superresponsive. The time of year was against me. Labor Day weekend had apparently already started.

That night I drove home with a sick feeling in my stomach, a kind of frustration that I hadn't experienced before. Usually I'm as excited as Nina and Layla about the start of school and am up for a fun long weekend, our last three days of summer together. I wanted to be an excited mom. But all I could think about were my other kids, my patients, and what might be happening to them. I felt like I was letting them down.

It was a hot summer afternoon, in the nineties, muggy. The kids of Flint would be playing outside, running through sprinklers. They would be swimming in the pool at the YMCA or University of Michigan–Flint rec center. Maybe they were sipping from a drinking fountain at Kearsley, a fifty-seven-acre city park that had been completely resuscitated in the last ten years with the help of some grant money and a master gardener and theater lover, Kay Kelly, who brought Shakespeare productions and children's plays there, along with a new bike path, a soccer field, and a new pavilion.

Nina and Layla were playing together when I got home, arranging stuffed animals in a make-believe zoo. They had amassed an unbelievable collection that included toys from my own childhood, now missing eyes and limbs, like Fifi and Sarah, my poodle and teddy bear. Nina was carefully facing all the animals in one direction so they could be attentive students in her classroom. Layla was dressing a stuffed dog in a ballerina tutu. Sometimes both girls played the same make-believe game, and the animals became their family. The father was a bear. The mother was a cat. The babies were ducks and lambs.

One day, a month or so back, the stuffed animals were all over the floor, in groups of two and four, multiplied. "What's going on, my little squid monkeys?" I asked.

"Parent-teacher conferences."

"Okay," I said, sitting on the floor, grabbing a stuffed animal, and joining in the playacting.

But not this weekend. When I saw them playing, my heart tugged to join them, but I was soon distracted by more thoughts of lead-tainted water.

The rest of the weekend was a blur—long, agonizing hours of pacing with my phone and being distracted. Elliott did his best to keep the girls occupied, but we could both see they were starting to notice. I was increasingly anxious and preoccupied. I reread reports about lead exposure, found more studies about water treatment, and looked again at the coverage of the D.C. crisis. On Tuesday morning, at the end of three days in purgatory, I sped to the hospital as if I were rushing headlong into battle.

ROEBUCK WROTE ME BACK, first thing. *Hallelujah.*

FROM: Dr. Michael Roebuck
TO: Mona Hanna-Attisha
SENT: Tuesday, September 8, 2015, 9:54 A.M.
SUBJECT: RE: report

Should have this by end of day today, depending on the complexity once we dive.

It didn't take the whole day. In a couple of hours, I had the blood-lead levels for my clinic patients. I instantly wrote to Kay Taylor, the director of our research department, with a request: "I need help with a data set. It's fairly urgent and fairly important—

lead levels for our patients in response to Flint water stuff. Let me know who can help me."

Over the weekend, I had come up with another possible way to get the surveillance data—not from the county but directly from the state health department. They had the blood data for all kids in Michigan. A year before, a nurse in the agency's Childhood Lead Poisoning Prevention Program who was in charge of "lead education" visited Hurley and had given a Grand Rounds one-hour lecture to our residents and faculty about lead poisoning and how to screen for it.

At my desk, I searched my email in-box and found her name, Karen Lishinski, MDHHS. And I saw that her email signature included a phone number. I was tired of not getting responses to emails, so I called her.

She answered the phone.

I couldn't believe it. A live human being.

After some quick small talk, I got to the point: "I'm concerned about lead. High levels are being found in some water testing. There should be some indication of this in the blood-lead levels. Has anyone looked at that?"

Matter-of-factly, without even a pause, she answered, "Yes, we looked at the lead levels over the summer, and we did see a spike."

They saw a spike? *Did she really just say that?*

My heart beat faster, but I tried to mask my excitement and horror with my doctor's deadpan. I asked, very politely, very calmly, if I could see those results.

"Sure," she said, "I'll send them to you."

"You have my email address?"

"Yes."

"Thanks."

And that was it. Suddenly I was on a roll. As soon as we got off the phone, I stared at my computer screen for ten minutes straight, and I don't think I blinked. A spike? *The state saw a spike?* Why wasn't this front-page news? I began doing breathing exercises, like the ones pregnant women are taught for childbirth. Inhale, then start counting. Exhale slowly. I took another breath and counted.

Nothing arrived after twenty minutes, then thirty. Perhaps she was still searching around the data docs and other reports in her computer, trying to find the summer lead levels. It wasn't as if she had to go to some underground storage vault like the one in *Indiana Jones* where the Ark of the Covenant was hidden.

To pass the time, I answered other emails. The array and amount of things going on would almost have been laughable if I hadn't been too stressed out about lead to notice. I was on a national committee that had revised online educational guidelines for residents, and there was a lot of back-and-forth about this. Our residency program had also just received a grant from the American Academy of Pediatrics (AAP) to establish a toxic-stress-intervention program for kids in the community—one of the few in the country. In the program we would teach positive parenting concepts to pediatric residents and Head Start teachers. National community pediatrics experts and AAP staff would visit our site in early October—three days of tours, meetings, and discussions. Lots to plan.

I made a little headway on those, then answered an email from a medical student in Cairo who wanted an interview, and another from a former medical student who was having a lonely and stressful transition to her residency at Children's National Medical Center in D.C., thanking me for a resident's "survival book" that I sent. Finally, there was an email from a resident who wanted to change her schedule because she was pregnant. (I tell my new residents that this is a perfect time to have babies—and we do our best to accommodate the schedules of new moms and dads in our program.)

My mom always said, "Mona, you're too busy." She complained that I pushed myself to the limit, signed up for too many boards and leadership positions, forged new programs, and took on extras. She wasn't the first to notice. Why did I do this to myself?

I had a sense that all my work was heading somewhere. It wasn't just a passion for training new pediatricians. It felt bigger than that, almost as if the more I did, the more I worked, the more creative and efficient I was, the more chances I'd have to make children's lives better.

But now, piling the work and weight of the Flint water issue onto the rest of what I normally take on threw off my sense of balance.

Meanwhile, the hours passed.

I kept clicking on my in-box and refreshing.

Nothing from the MDHHS.

Then Elliott wrote me. He had to go to the shoulder doctor for a cortisone shot—and hoped I could give him a ride.

I clicked on my in-box again.

Still nothing from the MDHHS.

Perhaps Karen Lishinski had been distracted by another call, or something else had come up, although I couldn't imagine anything, of any kind, anywhere, being more important than *high lead levels in drinking water.* How could a nurse at the state health department be unconcerned about a public health crisis? I was obviously finding it hard to think about anything else. Of course, it was possible that Nurse Lishinski was having some kind of terrible personal crisis in her own life.

Or maybe my request was perceived as above her pay grade.

Or below it?

Three hours later: *nada.*

I no longer cared about Karen Lishinski's workload, personal crises, or pay grade, or whatever else might be going on in the mysterious offices of the MDHHS. She needed to send that email. In my mounting restlessness, I turned my attention once again to the Genesee County Health Department. I needed a better answer from them. Something that was real and not gibberish.

Hoping to light a fire in their underwear with what the MDHHS's lead nurse had admitted about a "spike," I wrote to them again, for the third time.

And on a whim, I did something provocative and strategic. I added Dean Sienko's name to the list of email recipients, hoping to frighten them into action. The man was a major general, after all, and exuded integrity and leadership from every pore. They couldn't possibly ignore me now. Dean Dean had credibility and clout. I was proud of myself for coming up with this diabolical idea. I was launching a two-pronged attack.

FROM: Mona Hanna-Attisha
SENT: Tuesday, September 08, 2015, 2:30 P.M.
SUBJECT: RE: Emailing: Lead Prescription

FYI I just got off the phone with Karen Lishinski from the state's lead program and I asked her for Flint blood-lead levels. She said that in response to the concern about lead in Flint's water, they have already looked at the lead levels with their epidemiologist and it "appears" that there may be an increase, but they cannot comment on causality. I asked her to send me the raw data to take a look. I'll let you know if I get it. . . .

Lastly, I spoke with Dr. Dean Sienko about this issue last week. . . . [He is] with the MSU Public Health program in Flint now . . . [and] has offered additional MSU resources and his expertise as needed.

Mona

My strategic inclusion of Dean didn't go as planned, though. He was the only one to reply. Even worse, he hit "reply all."

To my knowledge, we never had a child with elevated lead where water was the principal or even minor concern; rather, it almost always came down to chipping lead paint.

Arrghhhh. The greatest obstacles to good science are assumptions and biases. Dean Dean was revealing his. I really liked him. And I couldn't really blame him. From the perspective of the public health community, the threat of lead exposure was connected exclusively to lead paint and paint dust, which were thought to be the gravest and most common threats. But particulate chunks of lead that break off inside a water pipe and find their way into a drinking glass or baby bottle can be just as loaded with lead as a chip of paint. Unfortunately, the dismissals and distortions of the CDC during the D.C. water crisis—and its refusal for years to admit that lead in water caused harm to children—had misled Dean Dean and many others in public health, just as Edwin Chadwick had persuaded a posse of do-gooders in the nineteenth century that miasma was spreading cholera.

It was bad enough to take on a county health department, but now I suspected that my battle might be much bigger, taking on preconceived notions and established dogma in the entire field of public health. It was incredible how little impact the D.C. crisis had. *You mean we need to worry about lead in water?*

By the next morning, when Karen Lishinski still hadn't written, I sent her an email reminder. I asked once again for the state blood data from the summer—and inquired if there was someone else at the MDHHS I should contact to help with this request. *Maybe she wants to shift responsibility elsewhere.* I was trying to think like a bureaucrat.

CHAPTER 9

Sit Down

MY FIRST VISIT TO FLINT WAS ON A CHILDHOOD TRIP TO AutoWorld, an amusement park that was one of the city's many revitalization schemes. Flint had once been a shining example of America's industrial prowess, a city known for the highest average income and lowest unemployment. But by the 1980s, Flint had fallen on hard times. Deindustrialization had hit, along with the oil crisis. The joblessness rate was the highest in the country, prompting the media to give Flint one of its awful but catchy monikers—the "unemployment capital of America." So when Auto-World opened in 1984, the promoters said it would draw one million visitors a year—and most important, employ 650 Flint residents.

As a loyal GM family, of course we went to AutoWorld. I was eight years old, so the Ferris wheel and carousel were my most vivid memories—and a bizarre ride that took my brother and me through a "humorous history of automobility" narrated by a talking horse. It wasn't very funny.

The park was planned as a way of honoring the automobile, as a testament to American prosperity, ethos, and mobile way of life. Amusement park rides were designed and built, along with tacky souvenir shops and fast-food restaurants, all with a car theme. The year the park opened, the Detroit Tigers won the World Series (a far, far bigger deal). James Blanchard, the governor at the time, heralded AutoWorld as "the rebirth of the great city of Flint."

And just like Disney World, AutoWorld had a cutesy Victorian Main Street with shops and eateries—in this case, a replica of Saginaw Street circa 1900 that included the Flint River. But the first attraction you saw, upon entering the theme park, was a history lesson: a rustic cabin and a terrible mannequin of Jacob Smith, the man who bought the land that became the city of Flint from the Ojibwa tribe in 1819. Interestingly, Smith married an Ojibwa woman and lived with her people as an Ojibwa himself. But the exhibit completely ignored the darker tale of the Treaty of 1819 and subsequent treaties that forced Native Americans off their land and pushed them into Canada.

I guess historic crimes and ethnic cleansing would have ruined the theme park vibe, so there was a lot about Flint that wasn't part of the show. The designers of AutoWorld wanted us to focus on the prospects ahead for Flint, not on the past. The sunny future was shown in an assembly line of robots, which should've been taken as a dire warning about the city's future, not a vision of utopia.

As it turned out, we got to AutoWorld just in time. In a matter of months, it turned out to be a gross disappointment, especially to its investors. Nobody showed up, except maybe my family. By December of its first year, AutoWorld was open only on weekends. By January, it closed completely. Over the next decade, it flickered on and off, opened and closed sporadically, each revival more pathetic than the last.

THE NEXT THING I remember about Flint was Michael Moore's career-making documentary, *Roger and Me*. It came out in 1989, when I was thirteen—just finishing eighth grade and living in Royal Oak. My family couldn't wait to see it—and we rented it from a video store as soon as it came out on VHS.

The movie, set in Flint, got a lot of attention, particularly in Michigan. Moore pioneered the social-justice gotcha documentary. And almost everyone got behind the way he went after Roger Smith, the CEO of GM, for closing the Flint plants.

My dad was a loyal GM engineer but also an old-school lefty, so he was torn about Moore's movie. He didn't want his company to look bad, but his sympathies were always with workers and workers' rights. As a professional engineer, his job at GM was outside the collective bargaining process, but he never lost sight of the fact that his wages and benefits—including his pension and gold-plated healthcare plan—were the result of the standards that had been set by the tough United Auto Workers (UAW) negotiations.

At thirteen, I picked up on his mixed feelings about *Roger and Me*—and came away with my own. Other than the seriously disturbing scene where a very cute rabbit is slaughtered on camera— beaten by a lead pipe—the most memorable part of *Roger and Me* for me was Moore's chronicling of AutoWorld: he showed how AutoWorld offered a rose-colored-glasses view of GM and the rise of the automobile. In so doing, he opened my eyes to how wishful thinking can be used to obscure the facts, leaving out inconvenient truths and lessons.

The years following Moore's movie were unkind to the city. The word *Flint* alone came to stand for the ravages of deindustrialization, what's left behind when a giant corporation abandons its birthplace and workers, outsources jobs abroad, and installs robots on the assembly line. By the time I got to the University of Michigan and studied environmental science deeply, I knew more about the tragedies of the auto industry: traffic accidents and fatalities, air pollution, global lead poisoning, dependence on fossil fuels that caused conflict around the globe—and even the way cars often encouraged Americans toward a soulless suburban existence. And global warming.

In medical school, I did my clinical training in Flint. I hadn't been there since AutoWorld, and it wasn't the failure stories that brought me. I was fed up with the bad press, the depressing stats, and the photos of industrial ruin. While the auto industry had truly used and discarded the city and cast it adrift, I knew there was so much more to it. And I knew I wanted to practice medicine there.

I came to Flint for its hope, but also for its lessons, both terrible and beautiful. Flint from its beginnings has been a place of extremes, where greed meets solidarity, where bigotry meets fairness, and where the struggle for equality has played out. Flint is where many people have been pushed down and many have risen. And where many have fought the good fight—and won.

LONG BEFORE CARS WERE made there, and as soon as the Ojibwa natives were unceremoniously driven off, Flint was a center of fur trading and lumber. This led, for logistical reasons, to the manufacturing of carriages, which led to the manufacturing of cars.

General Motors traces its roots to a carriage factory built in Flint in 1880—the Durant-Dort factory. It made its first successful foray into cars in 1908 with the opening of a Buick factory. As the demand for cars mushroomed, so did GM. Flint was home to GM's auto parts plants and the AC Spark Plug and Delphi Automotive Systems divisions. In 1913 a huge Chevy plant opened. Two Fisher

Body plants opened in Flint, by then a growing center of industry, and later GM built Buick City, a 235-acre state-of-the-art manufacturing complex, in northeastern Flint. At one time, GM's original Buick plant was the largest factory in the world.

The boom continued, fueled by World War II, when GM built tanks at a Grand Blanc plant just outside Flint. And south of the city, in the Willow Run Assembly near Ypsilanti, more bombers were built in a month than all of Japan built in a year.

Immigrants from Europe and the Middle East, from Poland to the Levant, moved to Michigan's industrial centers, including Flint, for jobs and a new life as wartime production expanded. Unemployment was 0.6 percent during the war. Women were given work, often clerical, but the war created opportunities for line jobs and higher pay, transforming America's workforce and society.

Between 1915 and 1960, more than six million African Americans, hoping to escape Jim Crow in the South, came north for those expanding employment opportunities—and during this Great Migration many were drawn to Flint, even though the hiring practices and working conditions of the auto industry were unjust. Black workers were segregated to the poorest-paying, most insecure, lowest-skilled jobs, often the dirtiest and most physically difficult. The high-paying trade jobs were usually off limits to them.

Antidiscrimination initiatives put in place by FDR and the UAW made it easier for black Americans to rise out of low-skilled jobs, despite fierce opposition by white workers, who felt threatened. By the war's end, no place on earth had more auto plants and auto employees—or a more racially diverse workforce—than Michigan's industrial cities. Ford Motor's centralized hiring tried to be color-blind, which gave it the highest rate of African-American employment in the country. But at GM, decisions were made by individual plant managers, which resulted in many fewer African Americans in their plants.

Housing was segregated in Flint, a situation created and enforced by many factors. Federal housing policies set up barriers of affordability and discrimination that made it difficult for African Americans to take part in the postwar housing boom. Restrictive racial covenants were legal and common, encouraged by GM. These covenants explicitly stated that "homes could not be leased to or occupied by any person or persons not wholly of the white or Caucasian race." A typical one said, "No negroes or persons of negro extraction (except while employed thereon as servants) shall occupy any of the land."

Real estate agents enforced the boundaries, and newspaper ads for apartments were unabashedly race-centric. Local banks, following federal loan guidelines, kept blacks in rentals, which were in limited supply and controlled by landlords looking to make as much money as possible for their unimproved apartments. Black families had to double up, sometimes triple up, in order to afford rental units. And that meant overcrowding. Only two residential areas of Flint were "designated" for blacks. One had notorious industrial pollution—it was literally adjacent to the Buick plant, where the housing stock was poor.

As historian Thomas J. Sugrue explains so well in *The Origins of the Urban Crisis,* a study of race and inequality in nearby Detroit, white violence also enforced racial segregation. The white riots of 1943—in reaction to the building of the Sojourner Truth Housing Project—left forty-three people dead. For blacks, the message was loud and clear: moving to a white neighborhood might endanger your life.

While schools were segregated, at least the ones in Flint were some of the best in the country. Charles Stewart Mott, GM's largest shareholder, controlled and funded Flint's public school system for some forty years, until the 1960s, through his foundation, which is still active today. Mott conceived of and created a top-rate community school system that integrated public schools, recreation programming, health and dental care, and children's camps. He envisioned a full-service one-stop system that cared for children in a comprehensive way. This innovative approach to community schooling heralded a public school golden age, an idea that gained mo-

SIGN POSTED AT THE SOJOURNER TRUTH HOUSING PROJECT, 1945

mentum and was eventually exported to three hundred school districts across the country.

OVER THE YEARS, I heard a little bit about Flint's labor history from my GM dad, but my brother taught me so much more. His passion for labor history, like his loud laugh, is infectious. As soon as he heard that I'd be doing my clinical training in Flint, he began sharing what he knew. In his eyes, the story of Flint is the story of social change in industrial America—where sit-down strikes and protests were organized, and where unions and management hashed out collective bargaining agreements that would guarantee workers better wages and safer workplaces—and dignity.

It was Mark's stories that gave me the idea of making sure all the new pediatric residents at Hurley started off their first year with a tour of the city during their Community Pediatrics rotation—to learn the neighborhoods and what made Flint Flint and how it forged such loyal and proud people.

The residents are driven along the main drag, under the "Vehicle City" sign that still adorns Saginaw Street. They see the remnants of the once massive Buick City and try to imagine a time when tens of thousands of workers clocked in and out. They also see Sitdowners Memorial Park, located near the corner of Atherton and Van Slyke, across from the current GM Truck Assembly plant. That's where they learn a different source of Flint's fame: workers coming together and taking action.

It happened almost as soon as GM began making cars. Fortunes were being made, but autoworkers were not given rights or paid for overtime, routinely put up with excessive heat, endured dangerous jobs, and were treated as if they were expendable and subhuman. Injuries on the factory lines—like lost limbs and fingers—meant no pay while they recovered and often no compensation for medical costs or permanent disability. Workers soon started to organize and demand higher wages and better conditions—as well as workplace safety measures. They organized a strike at the Studebaker plant in 1913 and another at Fisher Body in Flint in 1930.

The idea of a national union was a far-off dream until 1936. That's when things changed—and led to "The Strike Heard Around the World," or the epic Flint Sit-Down Strike.

A sit-down strike is exactly what it sounds like. Instead of mounting a conventional strike, picketing outside a workplace, the employees barricade themselves inside a plant and refuse to leave—stopping all production. It was a creative, radical, and subversive act, not to mention totally illegal.

And that's what happened in Flint on December 30, 1936, when workers occupied GM's Fisher Body Plant No. 1, demanding that the company recognize the UAW as their representative in negotiations. Retaliating, GM plotted to turn off the heat in the plant, in hopes of freezing the workers out.

The sit-down strike went on for a couple of relatively quiet weeks. Then on January 11, 1937, the guards at Fisher Body Plant No. 2 cut off the strikers' access to their food supply—and then, with guns and tear gas, attempted to force them to leave. They refused and fought back, throwing two-pound hinges at the plant guards and drenching the Flint police with water from firehoses. The event was later called the "Battle of the Running Bulls."

Another critical moment in the strike happened a couple of weeks later, on February 1, when Flint strikers creatively devised a plan to seize the giant Chevrolet No. 4 plant. Knowing that company spies would be undermining them, they tricked the company by announcing they would be striking at Chevrolet No. 9—and when GM fell for that, the strikers switched directions and occupied Chevrolet No. 4 instead.

This expansion of the strike effectively shut down GM's industrial output nationwide. Only after newly elected Michigan governor Frank Murphy refused to use force against the strikers—and even sent the National Guard to protect them—did President Franklin Roosevelt demand that GM recognize the UAW and end the forty-four-day strike.

My favorite union organizer has always been Genora Johnson Dollinger, a serious socialist who became a critical strategist during the Sit-Down Strike—and fundamental to its success. She organized the Women's Emergency Brigade and rallied women and children to sustain, support, and protect the strikers. She had to be physically dragged from her protests, which gave her the nickname the Joan of Arc of Labor. After being blacklisted in Flint, Genora kept organizing in Detroit—where she was a victim of a lead pipe attack—went on to work with the Michigan ACLU, and later led marches against the Vietnam War and helped found Women for Peace. Her tenacity and her ability to rally support for the labor movement helped make generations of workers' lives better.

What followed these labor victories was later called the Grand Bargain. In return for the workers' labor and loyalty, the company would pay them living wages and provide decent benefits. This Grand Bargain went on to encourage growth and the rise of immense prosperity, allowing workers to share in the consumer economy they helped create. The result was a thriving new middle class who could afford to buy the products of American industry, especially its cars.

THE CHILDREN OF FLINT SIT-DOWN STRIKERS, 1936–37

What began in Flint—a negotiated relationship between industry and the working class—went on to influence the wages, benefits, and working conditions of working people throughout America. It was in Flint that the middle class, and some would say the American Dream, was truly born.

THE MORE I LEARNED about Governor Frank Murphy, the more I loved him. Michigan has had its share of high-profile governors who went on to become national leaders, but nobody compares to Murphy. He was a true progressive reformer who was called a New Dealer before there was a New Deal. During his two years as governor (1937–39), Murphy defended autoworkers during the Sit-Down Strike and supported recognition of the UAW—during his term alone, the UAW grew from 30,000 to half a million members—and thereby risked his political future. Painted as a villain by congressional conservatives, Murphy drew the attention of the House

Un-American Activities Committee, which held one of its first meetings in Michigan to investigate the causes of the Sit-Down Strike. The committee blamed Murphy for Michigan's "state of anarchy," and Flint's city manager accused him of treason.

Murphy's vision of a just world caused him to lose his bid for reelection, but the strength of his convictions and leadership placed him in national positions almost immediately. As U.S. attorney general from 1939 to 1940, he established the civil liberties section of the criminal division of the Justice Department. And after FDR appointed him to the U.S. Supreme Court in 1940, Murphy championed outsiders, underdogs, minorities, and the underserved. In 1944, in his dissenting opinion in *Falbo v. United States*, he wrote lines that have inspired my belief in government as a force for good: "The law knows no finer hour than when it cuts through formal concepts and transitory emotions to protect unpopular citizens against discrimination and persecution." On behalf of another group of "unpopular citizens," he referred to the internment of Japanese Americans during World War II as the "legalization of racism" in his dissenting opinion in *Korematsu v. United States*. The first justice to ever use the word *racism* in a Supreme Court opinion, he went on to employ it seven more times before it disappeared from court opinions for two decades.

Throughout his short life—when he died at fifty-nine, thousands of newly organized workers and union members lined the streets of Detroit to pay their respects—Murphy showed tremendous courage. In 1925–26, when he was a judge, well before he achieved national recognition, he presided over what was billed as the trial of the century. A black physician in Detroit, Dr. Ossian Sweet, had dared to move his family from the black ghetto to an all-white working-class neighborhood. Racist whites attacked and invaded Sweet's home. One of the rioters, a white man, was killed in self-defense, and Dr. Sweet was charged with murder.

To defend him, the NAACP brought in the great trial lawyer and showman Clarence Darrow, famous for his defense of Eugene V. Debs in the 1894 Pullman railway strike. Judge Murphy, feeling that the trial would be an important test of justice and liberalism, assigned himself to the case, which led eventually to Dr. Sweet's acquittal.

A battle for justice was won, but the rest of the story isn't so happy. Dr. Sweet's notoriety led to the deaths of several family members and ended his career in medicine.

THE GRAND BARGAIN—AND THE PROSPERITY of the postwar years—expanded the middle class but also meant more housing stock was needed, especially those cookie-cutter starter homes that people wanted to live in. The suburbs became an easy answer. GM launched a strategy of acquiring large swaths of land in the emerging areas around Flint, where taxes were lower. It would build plants there, modern manufacturing facilities, and at the same time solve the housing crisis. The suburban exodus of manufacturing resulted in a suburban exodus of residents.

That exodus was mostly white, particularly during the fractious early days of school desegregation and busing. First, the Supreme Court outlawed racial housing covenants, and then came its 1954 *Brown v. Board of Education* decision desegregating schools. The momentous 1967 riots in Detroit and Flint seemed to scar whites with images of rebellion and destruction. Flint citizens passed a groundbreaking fair housing ordinance in February 1968 by forty-three votes—followed just a few months later by President Lyndon Johnson's federal Fair Housing Act. Together, these events left Flint reeling from the effects of extreme white flight: when an African-American family moved into a neighborhood—often in the face of intimidation and threats—white families moved out en masse, anxious that their property values would decrease.

Greedy real estate agents colluded to control neighborhoods in what came to be called "blockbusting." Real estate agents would go door to door, frightening whites with warnings of plummeting home values—and offering to put their homes on the market quickly, often below market prices. The agents then sold the homes to African Americans at a large markup above market prices. The scam continued at the banks, which charged blacks a higher interest rate. The upward mobility that white Americans took for granted in the second half of the twentieth century was denied to most blacks, but blockbusting—which continued into the 1970s—kept it out of reach even longer.

Worried about the city's future vitality and population, some city leaders, institutions, and entities, from GM to the UAW and the Mott Foundation, backed a 1968 referendum to bind the city and its suburbs to a regional government. The first effort had been made prior to white flight, as far back as 1958. Cities in the South and the West grow geographically by annexing suburbs, as in metropolitan Los Angeles, where suburban expansion didn't jeopardize the tax and economic base of downtown. But the residents in Flint's new suburbs, happily ensconced in their new houses and all-white neighborhoods, voted overwhelmingly against the referendum. The ramifications of this missed opportunity were grave and enduring. Flint was left isolated and abandoned, comprising only a bit more than its urban core.

Attempts to desegregate Michigan's public schools met with even more white voter anger—and violence—throughout the 1960s and '70s. The school busing wars raged, culminating in the bombing of school buses in Pontiac in 1971. Finally in 1974 the Supreme Court issued a decision that kids couldn't be bused across city lines. That decision, if it had gone the other way, could have integrated my hometown of Royal Oak with kids bused in from Detroit. Justice Thurgood Marshall wrote in his dissent to *Milliken v. Bradley* that the "Court's refusal to remedy separate and unequal education" leaves "little hope that our people will learn to live together and understand each other."

THE DOWNWARD TRENDS CONTINUED in Flint after the failure of AutoWorld in 1984. Leaders came and went. More GM plants closed. Many people left, seeking jobs and warmer climates. As an urban center, Flint has been in a man-made state of emergency for forty years.

Today pockets of old-money affluence remain in Flint, but you can also buy a house here for less than $30,000. And I'm talking about a decent house, similar to the Royal Oak home where I grew up. The national poverty rate is about 16 percent, but in Flint, almost 60 percent of children live in poverty, some in extreme poverty. Unemployment, unheard of in the city's glory days, is now epidemic. Violence has also taken a toll: there were 48 killings in Flint in 2015, including 42 gun deaths, and an additional 147 nonfatal shootings.

The population of Flint peaked at 200,000 but is now less than half that, which means thousands of abandoned homes, plummeting property values, and far less tax revenue. When the state of Michigan cut revenue sharing, the city starved even more. When I returned to Flint to work at Hurley in 2011, a new glimmer of revitalization was under way, with a few new restaurants, an enhanced downtown, and a new program of blight removal. But given all the city's challenges, it was just that—a glimmer.

Even as a girl I had guessed that places are more complicated and nuanced than a sentence or one explanation can offer. That has proven true. No single bad decision or unfortunate event created modern Flint. The greatest forces working against the city were racism and the corporate greed of GM, which pulled out of Flint, the city that birthed and nurtured it, to satisfy financial problems caused by a lack of imagination. But halfhearted, weakly enforced policies like desegregation, and the premature end of the Great Society, also played a role, along with deindustrialization and real estate greed.

What drew me to Flint—and has kept me here—was something else I saw. Something that made me think of my dad and mom, of the family stories I'd heard. It was the tenacity and endurance of its people, the passion of its strikers and workers, and its legacy of steel-plated grit and resilience.

WHEN RICK SNYDER DECIDED to run for governor in 2010, Flint was bleeding money, just as it was bleeding people and taxpayers. He campaigned as a moderate and a can-do guy. A former businessman, he had been the chief operating officer at now-defunct Gateway computers. When Acer acquired Gateway in 2007, he left the company and kept busy as a venture capitalist, selling a healthcare firm for $200 million to Johnson & Johnson.

With his practicality and financial know-how, Snyder vowed to make Michigan government work like a business. While his focus on metrics and calls for "relentless positive action" always sounded creepy and vaguely cultish to me, there was definitely hope when he was elected that he might join the ranks of other successful Republican moderates, like George Romney and William Milliken, who'd governed Michigan fairly and successfully.

But almost from the start, a sinking sense of disappointment in Snyder set in. He wasn't a moderate at all—or at least he didn't govern like one. Almost immediately the Tea Party started pushing him to the right, and then he just kept going in that direction.

Governor Snyder's enthusiasm for the emergency manager law seemed totally genuine and unforced—not something he was pushed to do by the Tea Party or anybody else. One of the first bills he signed was Public Act 4, a beefed-up version of a prior EM law that allowed him to take over a municipality that was beset with financial problems. Even though Michigan voters shot down Public Act 4 in a referendum, the ideologues in the legislature ignored the voters and passed the law again, one month later.

The appointed EM had one job: extreme austerity. (Someone called Snyder's EM scheme "austerity by dictator.") The EM had to balance the books, no matter the cost. The EM law has so many miserable aspects, it's hard to know where to start. City employees and their pensions and benefits were a primary target.

The extreme austerity—which was also behind the water switch—

decimated Flint's police force. Between 2008 and 2016, it shrank from 265 sworn personnel to 98.

Selling off city resources was also part of the grand plan. The EM law also allowed for the privatization of public institutions, even possibly Hurley Medical Center. By the time I got deep into the water crisis, three different Flint EMs had come and gone, each lasting only a short while.

It didn't really matter who they were anyway. The governor's office in Lansing was calling the shots.

—

CHAPTER 10

Jenny + the Data

THE LACK OF RESPONSE FROM THE STATE, SPECIFICALLY THE health department, sent me into a dark hole. I just didn't get it.

Just as I began to fall into despair, Elin wrote to me with good news. For the last week, she had been working behind the scenes quietly, like a spider spinning a web. The first strand in her trap was marshaling political support. Elin was working as a water policy analyst for a Washington-based nonpartisan think tank, the Northeast-Midwest Institute, where she was finishing up a massive report on fracking, a project that had kept her in regular communication with members of Congress, including the office of Dan Kildee, who represented Flint. She got in touch with Kildee's office to share her concerns about Flint water and passed along my name and number to a staff member.

Elin's view all along was that the D.C. crisis had been prolonged and worsened because the District lacked congressional representation. For all Flint's problems, we at least had congressional representation. If Representative Kildee could be brought in, made aware of the lead levels in Flint, we'd have more clout and more options.

I didn't know much about Kildee, except that he'd succeeded his uncle, Dale Kildee, who had been the congressman for the Flint area for almost four decades and had been an old-school New

Dealer. Even if the nephew had gotten into office by trading on his locally famous name and had nothing else going for him, any congressman who represented and lived in Flint would surely have to help. The controversy over the water switch had been bubbling for a while now. Some of his constituents were definitely mad, and hopefully they were also contacting his office with concerns.

Sure enough, after a brief email exchange, Jordan Dickerson, Kildee's legislative director on Capitol Hill, called me to talk. He sounded young, earnest, and thanks to Elin, already well-informed. He knew about the Del Toral memo, the pushback from the EPA, the lack of a response from the state agencies, and that I was looking at my kids' blood-lead levels. He asked what I knew so far and what I needed.

Quickly, I spelled out my request. The county and state collected children's blood-lead levels for all of Flint. Could Representative Kildee's office help me get them? To do the science right, we needed that large sample.

After so much silence and feeling like I was shouting into the wind, I was greatly relieved to hear excitement in Jordan's voice. He was on board and had the complete support of his boss. Better yet, he couldn't wait to get started.

ELIN SENT ALONG NEWS of another big development: Marc Edwards and his Virginia Tech research team had more results posted online from extensive citizen water testing in Flint. In case there was any doubt about the urgency of the city's water situation, the headline of Edwards's report was bold and in all caps. If it were spoken, it would have been yelling:

"FLINT HAS A VERY SERIOUS LEAD IN WATER PROBLEM."

The numbers were pretty dramatic. "Forty percent of the first draw samples are over five parts per billion. That is, 101 out of 252 water samples from Flint homes had first-draw lead more than

5 ppb." The report went on to say that "several samples exceeded 100 ppb, and one sample that was collected after flushing exceeded 1000 ppb." No level of lead is actually safe, but this meant a significant number of households had levels above the federal action level of 15 ppb.

Those are scary numbers, sure to shock anybody with any knowledge of the consequences of lead as a neurotoxin. "Until further notice," Edwards recommended that Flint water be used for cooking and drinking only if filtered first or after the tap was run for a full five minutes at a high flow rate.

Elin and I texted back and forth, horrified by the numbers and wondering how the state would react. *Can they really keep ignoring this?* Both Dean Dean and Elin wrote with the news, according to *MLive,* that Edwards was returning to Flint the following week and was scheduled to appear at a town meeting on the evening of September 15, at Saints of God Church on Forest Hill Avenue. I added the town meeting to the calendar for my Community Pediatrics residents.

Then I wrote to Elin, "I think we should meet him."

She called immediately, something she did only when she had too much to say and didn't have the patience to type out her response. "All the water people are afraid of him," she said.

"Should I be?"

Elin used words like "too intense" and "radioactive" to describe Edwards, saying that he didn't play nice and seemed willing to attack almost anyone. If we wanted to convince state authorities to get on the right side of this issue, any connection to him could backfire. His battles with the EPA, the CDC, WASA, and the D.C. government were notorious in water circles and had sparked continuing investigations, hearings, and court cases. It had been a long and painful decade fighting governments and water utilities to root out lead from their water, an uphill battle that had left Edwards, according to things Elin had heard, emotionally and psychologically damaged.

I googled Edwards while Elin and I talked. Wow, he even had a

Wikipedia page. As soon as she and I said goodbye, I dug into it thoroughly. The page described the truly awful story of the D.C. crisis and listed Edwards's various papers, awards, and accolades. I wasn't really sure what a MacArthur Fellow was, so I clicked on the link to that website and found the page that described the genius award he'd received in 2007.

> Marc Edwards, a civil engineer, is playing a vital role in ensuring the safety of drinking water and in exposing deteriorating water-delivery infrastructure in America's largest cities. An expert in the chemistry and toxicity of urban water supplies in the United States, he has made significant advancements in a broad array of areas, including arsenic removal, coagulation of natural organic material, and the causes and control of copper and lead corrosion in new and aging distribution systems. Melding rigorous science, concern for public safety, and dogged investigation, Edwards's recent work focused on the identification and analysis of lead contamination in the Washington, D.C., area's local water supply.

I discovered that money came with a MacArthur award and wondered if Edwards had been able to pay off the second mortgage that Elin said he'd taken out to cover the costs of his own research in D.C. Who was paying for his work in Flint? It had to be pricey to travel five hundred miles with all these grad students time and time again.

A few photos were offered at the bottom of his MacArthur Foundation page.

There he was, Scary Marc. Except he didn't look radioactive. Angular and chiseled, the guy was seriously handsome, as if he should be starring in a movie about water, not testing it himself.

I couldn't help but notice the bizarre necktie he was wearing. It was way too wide and had a couple of baby panda bears with really sad-looking faces on it. Who was dressing this man? Was he for

MARC EDWARDS,
IN THE PHOTO
I SAW ON THE
MACARTHUR
FOUNDATION
WEBPAGE

real? How could you take a man in a panda tie seriously, much less find him scary? But apparently everybody in the water world did.

I FOCUSED AGAIN ON the blood-lead data from my clinic. Kay Taylor, the head of Hurley research, had answered my frantic email within an hour, with an equally urgent reply. Bless her. Kay has always been so supportive—and she epitomizes everything I love about working at Hurley, such a passionate, altruistic, and lean place to treat patients and train new doctors.

Compared to other teaching hospitals, our research department isn't that big—just a handful of staff—but we're nimble and scrappy and always evolving to meet the research needs of my residents and faculty.

Kay assigned a research coordinator, Jenny LaChance, to help me. Jenny and I had previously worked together on tight deadlines for grant applications and research projects, and had known each other since I was a medical student at Hurley. We were a good team. I love crunching numbers and trying to harness the power of data, but Jenny took that passion to a whole new level. Smart, logical, superscientific, and verging on obsessive, she loves nothing more than designing a foolproof study and figuring out the best way to collect and analyze data. Jenny could talk your ear off about power and sample size, positive predictive value, statistical significance, and absolute risk reduction, among other esoteric subjects.

I left work early that day to take Elliott for another opinion about his shoulder. Along the way, as I weaved through Metro Detroit traffic, we talked mostly about the girls. They were in the middle of their first week of school and settling in fine. Homework was light. There was soccer practice, but no games yet. Other parents were helping with carpooling, and Elliott's parents were bringing over the occasional meal. As we got deeper into the logistics of soccer and meals, I mentally checked out—thinking about a dozen other lead-related things. A pang of guilt brought me back to the car, and I forced myself to think about home life again.

In the waiting room, I tried to catch up on the glut of half-ignored emails that were taking over my in-box—questions from my pediatric residents, scheduling dates for nutrition education meetings, motivational interview training, following up on a new resident from India who'd gotten in a car accident just a day after getting her driver's license, and more back-and-forth banter with Roebuck about the new software tool that the IT folks were embedding in our EMR to help residents safely hand over patients between shifts. ("Your peeps are awesome, give them a hug for me," I wrote to Roebuck. He replied: "IT nerds don't hug. Makes them uncomfortable. Instead we spread out fingers and either say 'live long and prosper' or 'nanu nanu.'")

I CALLED AN URGENT meeting for the next morning, September 9, with Jenny and my team of Community Pediatrics residents. I explained my hypothesis: that lead in the Flint tap water had ended up in the bodies of our children. Assembled around the rows of tables and chairs in our pediatric classroom, adjacent to my office, we pulled up the raw labs—the unfiltered data from patients' charts—on the conference room screen.

A quick glance at the numbers on the screen—before and after the water switch—appeared to show an increase. But I couldn't give just a quick look and be certain, the way I might quickly look at the itchy bumps on a young patient's legs and the miserable look in her eyes and immediately identify a rash as scabies—*Look, here are the classic signs, the pathognomonic burrows.*

Analyzing data requires a nuanced approach and much more time. As I had learned as an undergraduate and later in medical school, a good study needs to take into consideration an array of factors to get the science right, then must be carefully designed for accuracy and clarity. Numbers may appear to be black-and-white and simple, but they have a way of hiding secrets if a study and analysis aren't properly done. A lot of thinking goes into a study, and often there are many revisions, times when you have to start all over from scratch.

In our meeting, we discussed the kinds of questions that researchers have: how to collect information, how to organize or group the data, and what should be included or excluded. For instance, we had blood-lead levels for children of different ages. Should we lump them all together or limit the age range to children under five, since they are the most developmentally vulnerable?

As we looked at the numbers, we realized we had no idea where these kids lived. They had come to our Flint clinic, but did they even get Flint water? Jenny and I searched on ZipCode.org for

Flint and decided to use 48501 through 48507 on our data set. I wrote to Elin to ask if the water supply went just to Flint or beyond the boundaries of the city. Then I wrote back to IT and asked if they could run the numbers again, including zip codes in the report.

I heard back just a few minutes later. That's the miracle of electronic medical records. So much time goes into crunching—or into fine-tuning and designing a study. But once you ask the right questions, spitting out the results happens in an instant.

We were far from finished with the fine-tuning process. We still had other things to sort out, like how to filter out repeats—patients who were in our system twice because their lead levels were measured twice in our time period. And when we saw a superhigh lead level reported, should we look individually at each of those patient charts to see what the determined cause was? If we saw that the child had eaten paint chips, should we exclude them from the study?

We knew we'd be doing the research twice, for data sets on lead levels before the water switch and after. But how were we going to break up the time periods? Should we look at a date immediately after the water switch, or should there be a lag time for the leaching of lead? How long did it take for the leaching to occur?

The residents began pulling up research on their laptops. As I spoke, their focus on me was intense, and I could see their minds were fully engaged. Most of the research that medical students and residents do is important but academic: one special report of a rare case of encephalitis, or a study of patterns of treatments that happened years ago, in other places, other hospitals. But here we were, talking about lead poisoning, lead exposure, and blood-lead levels, in numbers from the previous year and over the summer, numbers from just a couple of days ago. It was real, something that was happening all around us, the blood of our own patients, and water that flowed in the pipes of our own city, where we sat. The residents were engaged in a way I'd rarely seen before, vibrating with a weird new energy, tense but invigorated by the feeling that we were finally *doing something*. And our results weren't going to be stuffed away in a digital archive and forgotten. Our results could change our world.

The residents eagerly pulled up related literature and the electronic charts of some of the kids with the highest lead levels: "Oh, this one says there was peeling paint in their house," or "This one doesn't mention a potential source." Another said, "This kid was shot." Bullets left lead in the bloodstream too. We decided to look only at kids five and younger; to look at only one level per child—the highest level; and to include all kids no matter what was in their chart as a potential source. As for time period, we agreed that we'd look at an equal length of time before and after the water switch—January 1, 2013, to April 24, 2014, versus April 25, 2014, to September 9, 2015.

Jenny took careful notes on her laptop while we talked and said she'd have something soon, even by tomorrow. I knew from our other projects together that, like me, Jenny wasn't a nine-to-fiver. Once she got going on something, she took it home and kept working. Putting a study together was like working on a jigsaw puzzle—it was kind of addictive, hard to stop, and offered a dopamine hit of pleasure when you reached the end.

When the meeting finished, the residents filtered out of the room to see patients in the clinic. They were silent, deep in thought. I had tasked them with overnight assignments—looking up more literature on lead exposure, finding similar studies, taking a deeper dive in a number of areas we needed to know more about. They knew this work loomed large for me, for the clinic, and for all of us, but especially for our patients and our community. We would be fashioning an argument out of blood tests, numbers, and zip codes.

When the last resident left, I noticed Jenny hanging back, waiting to talk to me. She had a few questions, she said, and I could tell they were personal. She had a look on her face that wasn't about numbers. It's the same look I see on every parent's face who comes to the clinic worried about her kids.

It was about Drew, her baby son. He was just a year old and still nursing. Jenny wanted to know, if she was drinking Flint tap water at the hospital, could she be passing along lead to him in her breast milk? She was using a breast pump and sending bottles of it along to his daycare in Grand Blanc, a suburb south of Flint.

I knew this was the kind of question all the parents and caregivers in Flint would have once they knew what we suspected—including all the moms who worked with us at the clinic.

"Aside from coffee and tea, or whatever I get at the cafeteria," Jenny said, "I haven't been drinking the water at work." As a Flint native, she had been alarmed like everybody else when the city's water source was switched to the river. And when she saw the early news reports of water-quality issues, she had the luxury of taking extra precautions, even bringing bottles of tap water from her home, which was outside Flint. I assured her that all those protective steps were good—and she should keep doing them.

"But what about when I was pregnant with Drew?" she asked. I could see her calculating the date of the water switch in relation to her pregnancy. Drew was born in August 2014, five months after the Flint River started flowing out of the tap. A look of worry and sadness passed across her face. "That means seventy-five percent of the time I was pregnant, we were on that water. And I was coming to work every day and drinking the water."

"I know," I said, nodding, and then reached out to comfort her.

Jenny studied my face as much as I was studying hers.

"Do you think Drew's okay?" she asked. "Is there anything else I should be doing? He's been sick a lot lately." Drew was due for a bunch of vaccines that week, she told me, but was still recovering from his first ear infection. Jenny, being an obsessive researcher, had become paranoid about antibiotics, vaccines, the germy environment of daycare, and all the other worries that can plague a parent, especially a parent let loose on the Internet. So much of what we do as pediatricians is listening, reaffirming, and reassuring, especially first-time moms and dads.

"He'll be fine. The ear infection is a common virus. Don't worry.

Daycare is good for him. All the germs will make him stronger," I told her. "He does not need antibiotics. And vaccines are important."

After a parade of reassurances and a tight hug, we said goodbye, and I sat alone in my office chair for a few minutes, trying to clear my head. I felt bad for Jenny and knew how terrible it was to have mom-worries. I remembered those days. But imagine how the moms of Flint were going to feel when they did the math for themselves. And think of the ones who were pregnant, like Jenny, and were not just having an occasional cup of coffee at work but were drinking the water all day long. How worried and angry they'd be. Who would answer to them?

EARLY THE NEXT MORNING, on September 10, Jenny texted me. True to her hardworking (and somewhat obsessive) style, she had already run the numbers—and had completed a preliminary analysis of blood-lead levels from Hurley Children's Clinic.

JENNY: I got something.

ME: Email it.

JENNY: Okay. I'm coming to your office.

I was sitting at my office desk when her PDF came in with the actual statistical analysis. It was just one page. Easy to read and comprehend.

Comparing children under five years of age with elevated lead levels (greater than or equal to 5 μg/dl) for about the same duration of time pre–water switch (January 1, 2013, to April 24, 2014) to post–water switch (April 25, 2014, to September 9, 2015), the percentage increased from 1.5 percent to 8.5 percent.

The p value was 0.007.

In statistics, the "p value" shows statistical significance. And the lower the p value, the stronger the significance.

Anything less than 0.05 is significant. There it was.

Lead. It was real.

The impact was there in the numbers.

It was exactly what Elin had feared, and what Del Toral had suspected.

My clinic results showed more kids with higher lead levels since the water switch in 2014. The increase in the percentage was there, and it was big and statistically significant. But the sample was so small—4 out of 270 kids in the pre-period who tested high, and 6 out of 71 kids in the post-period.

Jenny appeared in my office, out of breath. She'd sprinted down three flights of stairs in a hurry. Even though her cheeks were flushed, her skin looked unusually pale, and there were dark circles under her eyes. She was tired and probably hadn't slept much, having run numbers all night.

We didn't even say hi or greet each other. We jumped in immediately, talking about holes in the analysis—or potential holes. One way to produce a perfect, unassailable study is to be a scientific devil's advocate. It's the way clinical reasoning works. Also, part of me was desperately hoping to find an alternative explanation for the results we were getting. I didn't want them to be true.

For instance, the percentage of kids with elevated lead levels was higher and was significant. But there could be a practical explanation. What if the six kids with high lead levels were siblings in a family that lived in one home that was infested with lead paint?

And we should have more test results, or data points. Where were they? Why were so few kids tested after the water switch compared to before the water switch? Before the switch, 270 screenings had been done. After the switch, only 71.

What was going on?

What I needed was a study that left no room for doubt. I needed a larger sample—and that was going to require getting records from the county or state. After I sent more persistent emails, the county health department finally said they could retrieve lead levels only one at a time—by downloading a PDF for each child. Was this how they did basic public health surveillance? (It was hard to imagine how these public servants slept at night.) Of course, Kildee's office might come through, but it could take time. There could be delays.

"We need more data, Jenny. We need more—"

"Wait a second. Hurley pretty much processes the labs for the entire county, not just for our clinic. Couldn't you just pull the numbers pretty quickly with our EMR?"

"I will email IT right now."

"Hold on," Jenny said. "I think we'll need to get an IRB approval for that."

She was right, we did. IRB stands for institutional review board. Since I was the director of the residency program at Hurley, I had the authority to review data from my own patients and patients at the Hurley clinic, but I didn't have the power to look at data from patients outside our practice. To get access to a much bigger sample, Jenny and I would have to have a review board at Hurley give us permission.

All research done on humans is protected by ethical guidelines that were established in 1979 in reaction to terrible historic incidents in which people were experimented on in cruel and exploitative ways, especially poor, disadvantaged, or mentally ill people. The infamous Tuskegee experiment that I teach my residents about—when government researchers withheld treatment from black subjects with syphilis—is a prime example of what an IRB is meant to keep from ever happening again. So while the process of getting an IRB approval could be bureaucratic and time-consuming, the principle behind it—that people are more important than any institution or system or research project—was so good, solid, and right, I couldn't complain.

At a big university, it could take months for researchers to get approval for a study; those big institutional review boards sometimes met only four times a year. One of the blessings of working at a small teaching hospital was how quickly we could get things done. At least we hoped.

"I'll give a heads-up to the IRB folks," Jenny said. She worked just a couple of doors from their office.

And if the stars were aligned and we busted our butts—and I really pushed it—we might get an IRB approval in a week or two.

MEANWHILE BRAD WURFEL, the rabid pit bull of MDEQ, was going after Marc Edwards. After the findings of the Virginia Tech study were posted, the local media picked up the story, which prompted MDEQ's vicious pushback. It was all over the morning news. Wurfel criticized Edwards for "fanning political flames irresponsibly" and described his Virginia Tech team as a group that "specializes in looking for high-lead problems."

Trash-talking the leading water expert in the country is a stupid strategy, if you think about it. Wurfel told *The Flint Journal,* "They pull that rabbit out of that hat everywhere they go. Nobody should be surprised when the rabbit comes out of the hat, even if they can't figure out how it is done."

How long had it taken Wurfel to polish these cute quips? If only MDEQ had spent as much time listening to the people's concerns and carefully looking at Edwards's findings as they had minimizing and dismissing them.

But now that I was lying awake all night, eyes open, mind buzzing, I used the time to read all the books I'd ordered on the public health history of lead. I was beginning to see that over the last century, lead seemed to bring out the very best and the very worst in people. And I knew already which side of history Wurfel and his MDEQ bosses would be on.

—

CHAPTER II

Public Health Enemy #1

I N Flint, everybody knows the name Kettering.

His inspiring quotes can be found on plaques and signs around the city: "Believe and act as if it were impossible to fail," or, my personal favorite, "My interest is in the future, because I am going to spend the rest of my life there."

Our world-renowned engineering school in Flint, once called the General Motors Institute, is now named after Kettering—as are a number of honors and prizes given out in the city. There is a city near Dayton, Ohio, population 56,000, that was named after him too.

It makes me a little sick now to remember how proud we were, and how much we celebrated in 1992, when my dad received a "Boss" Kettering Award, GM's highest engineering award.

Good ol' Boss Kettering. GM has tried to keep his flame alive. He was an engineer, an inventor, and the head of GM's research department for twenty-seven years, from 1920 to 1947. He was on the cover of *Time* magazine in 1933. By most accounts, he was jocular and adventurous, loved by his family and friends—and his breakthroughs in automotive technology, including the invention of the modern car starter and the two-stroke diesel engine, made a lot of money for GM. Along with Alfred Sloan, the president of

GM at the time, he helped to found Memorial Sloan Kettering Cancer Center in New York—and personally promised to oversee the organization of a cancer research program employing the most cutting-edge industrial techniques.

But sometimes you have to look away from fanfare and familial love, corporate achievements, and even remarkable philanthropic work and look at a person's entire legacy. Even I didn't know much about the man until I started doing more reading. I wasn't sleeping anyway.

To my mind, there is no greater public health villain.

LEAD IS PROBABLY THE most widely studied neurotoxin, a metal that humans—including my ancient Assyrian ancestors—have been using for thousands of years. Even the word "plumbing" is derived from the Latin word for lead, *plumbum*. On the periodic table of elements, the symbol is Pb. Its wonder was its remarkable malleability. The Romans used it to coat their famous aqueducts two thousand years ago. It was also employed in almost every aspect of wine-making, something else the Romans were famous for. A food additive, it was sprinkled on everything, like salt. All this lead in the Romans' food, wine, and water has provoked theories that lead poisoning contributed to the fall of their empire.

Since the first humans used lead, millions of tons of it have been dug from the earth and dispersed into the environment—far more than mercury or arsenic, the other two toxic trace elements. Aside from plumbing, its most widespread preindustrial use was in paint—to make it washable, brighter, and much more durable. Art historians suspect lead pigments may have contributed to the illness and demise of many great painters, including Correggio, Raphael, and Goya. There's even a story that Vincent van Gogh was fond of the flavor of one pigment in particular and liked to suck on his paintbrushes. That wouldn't surprise a pediatrician. White lead is noted for its sweetness, which is why lead paint tastes good to kids.

Although lead was known to be poisonous and unhealthy even in ancient times, its convenience—particularly the durability of lead paint—created a conflict between public health advocates and industry that pretty much continued into the twentieth century, as Christian Warren describes so well in *Brush with Death*.

Everybody knew lead was toxic, but what it did to the human body was insidious and invisible, while its benefit to industry was tangible and quantifiable in dollars. The story of lead in the United States, where industry has the upper hand, is a little different than in Europe. After childhood poisoning was linked to lead paint in 1904, several European countries banned its indoor use in 1909. But when the League of Nations banned lead paint in 1922, the United States declined to go along. At that point, the regulation of lead paint in this country was almost nonexistent. Lead was big business in America.

Given the rejection of lead in 1922 by most of the world, it's kind of amazing that just a year later it was added to gasoline in America. It was heralded as a giant technological breakthrough, a "gift from God." It's funny how we talk about "unleaded" gasoline now, as if lead were a natural component of gas and removing it required a complicated chemical process, maybe even at great expense.

The entire concept of "unleaded" is a misnomer, in fact. Lead was mined from the earth and added to gasoline. The reason? To stop engine knocking, an annoying noise in early car engines that occurred when the mix of fuel and air burned unevenly. According to GM legend, it took Kettering a decade of intense experimentation to find his antiknock compound. He had alternatives, like ethanol, but it could not be patented and profited from, unlike tetraethyl lead (TEL), for which GM obtained the patent. The health hazard of TEL, over and above normal lead, was no secret. Fat-soluble and easily absorbed by the skin, TEL was even tested by the U.S. War Department as a nerve gas. Just five teaspoons applied to healthy skin could be fatal.

Five teaspoons on the skin . . . this didn't stop GM or even slow the company down, it seemed. It collaborated with DuPont and

Standard Oil/Exxon on a new "high octane" style of gas, dubbed "ethyl," that allowed cars to drive better and engines to work more efficiently, never to knock or ping.

As soon as the coming of a new "improved" gasoline was announced, public health experts began writing about the dangers of lead. An anti-lead movement rose up and gained ground, as always seems to be case, only in the aftermath of a tragedy. In October 1924, noxious fumes poisoned and killed workers in a section of a Standard Oil refinery in New Jersey where the new gasoline with TEL was being made. Several exposed workers were straitjacketed after exhibiting paranoid behavior, experiencing delusions, and becoming violent. When the first death occurred—seventeen people died in all—the local district attorney called for an investigation, which launched a media frenzy. The tabloids referred to the lead fumes as "loony gas."

For the remainder of 1924 and most of 1925, the public health debate about tetraethyl lead in gasoline was one of the sharpest of the early century—much more heated than what eventually transpired over lead in paint, or than what has yet to transpire over lead in plumbing.

Now I get to my favorite (and the only uplifting) part of this story. It's about Alice Hamilton, who lived in the time of Kettering—and fought him with everything she had. Hamilton was a social justice pioneer, medical doctor, and professor who had specialized in lead toxicology cases among factory workers before turning her attention to GM. She pushed the company to find a better, safer alternative to TEL—and insisted that the introduction of lead on a widespread basis would have a catastrophic impact on public health.

"I would like to make a plea to the chemists to find something else," she said at one public hearing. "I am utterly unwilling to believe that the only substance which can be used to take the knock out of a gasoline engine is tetraethyl lead." She was right—there were alternatives. But public health wasn't foremost in the corporate mentality; profits were.

Hamilton is now one of my heroes, a stubborn badass who devoted her life to improving the lives of workers, the poor, and children. When things became stressful for me in Flint—and there were some tough days to come—I thought of how hard she tried to right a wrong, to make the world safer for kids, and to prevent a totally preventable public health scourge. Hamilton was a visionary who studied medicine at the University of Michigan at a time when there were few women in the field. In her memoir, she said that her studies in Ann Arbor had been transformational. (I could say the same thing about my time as an undergrad there.) Michigan gave Hamilton "my first taste of emancipation," she said, "and I loved it." Ditto that.

From there, Hamilton studied bacteriology and pathology in Germany from 1895 to 1897, just a few years before Paul Shekwana was in England, and she returned to the United States to continue her postgraduate studies at the Johns Hopkins University School of Medicine. Then she moved to Chicago to become a professor of pathology at the Woman's Medical School of Northwestern University.

In Chicago, she also became a member and resident of Hull

ALICE HAMILTON,
AT HER MEDICAL
SCHOOL GRADUA-
TION, 1893

House, the first settlement established by Jane Addams and Ellen Gates Starr, where newly arrived and impoverished European immigrants were given full or "wrap-around" services—from healthcare to education and advocacy—and a safe place to live. Hamilton opened a well-baby clinic and eventually expanded care to include children up to eight years old. While at Hull House, she began touring city slums, mineshafts, and factories, researching industrial diseases and occupational hazards. She discovered more than seventy industrial processes and jobs that exposed workers to lead.

When she was offered a position at Harvard Medical School in 1919, she was the foremost American authority on lead poisoning, and one of only a handful of specialists in industrial disease. But when Harvard University offered to hire her as its first woman professor, it made three stipulations:

1. She was forbidden to enter the Faculty Club.
2. She could not have tickets to football games.
3. She was not allowed to march in the commencement procession.

She took the job anyway. It's a lesson in strategic thinking, something she was especially good at. During Hamilton's years fighting lead in gasoline, she was adept at feeding tragic stories of lead poisoning to the popular media, knowing this would light fires that no academic study could.

Between 1924 and 1926, the lead-in-gasoline controversy was debated in 124 daily newspaper articles in the New York City dailies that were picked up in hundreds of wire-service stories around the country. When GM and Standard Oil pushed back, they claimed that Hamilton and other anti-lead activists were "hysterical"—that code word for "stupid woman, overreacting"—while the industry's work to improve gasoline was based purely on rational science.

The main medical expert defending GM's use of lead was its own in-house toxicologist, Dr. Robert Kehoe. At a conference called by Surgeon General Hugh Smith Cumming to resolve the conflict in May 1925, Kettering told his dramatic story of how he had experimented with dozens of additives and eventually discovered the wonders of TEL. As Alice Hamilton and other public health advocates looked on in horror and disgust, this mendacious testimony was followed by a succession of industry reps and medical experts backing up Kehoe's claims that TEL was totally safe and that there was no suitable alternative. He insisted that lead was naturally occurring in the environment and could be processed and eliminated by the human body without any health risk or danger.

During a break in the conference, Hamilton is supposed to have confronted Kettering in a hallway and said, "You're nothing but a murderer!" I'm not sure this story is true; it doesn't really go with her style, which was stubborn and persistent but always polite. But I can only imagine how angry she might have been that day, face-to-face with such a liar.

To conclude his testimony about the safety of ethyl, Kehoe made a diabolical offer: GM would discontinue use of TEL immediately if it could be proven to be harmful. This established a new precedent—later used by the asbestos and tobacco industries for decades—that forced public health advocates to prove harm before action could be taken. It goes against prudence and any sort of good judgment to allow dangerous practices to continue until enough evidence mounts to stop them—putting lives, even the planet, at risk. It was a turning point, not just in the history of lead but in so many other aspects of public welfare and safety. "Safe until proven dangerous" became known as Kehoe's Paradigm, or the Kehoe Rule. The approach was later taken by climate change deniers.

The public health approach, however, is far wiser than the Kehoe Rule. The Precautionary Principle holds that a product or chemical should be considered unsafe unless the manufacturer can prove otherwise.

To reach a decision in 1925, the surgeon general called a small committee to investigate the hard evidence on leaded gasoline—and to test workers for lead levels in their bodies. In seven months of study, the committee found that those who had worked with ethyl gasoline did have slightly elevated levels but none demonstrated clinical poisoning. It determined that TEL could not be proven to be harmful. Kehoe won. TEL went on the market.

I can imagine how devastated Alice Hamilton must have been, and how she must have despaired for the future, while Kettering celebrated his victory. The use of lead continued. Leaded gasoline, or "ethyl," became phenomenally successful, appearing in virtually all major-brand gasoline sold in the United States and around the world throughout the ensuing three decades. The costs of the surgeon general's decision have been incalculable. By 1960, almost 90 percent of all gasoline contained TEL. By 1965, according to the work of geochemist Clair Patterson at Caltech—who inadvertently stumbled onto the man-made contamination of lead in the environment while on a quest to determine the age of the earth—the average American had one hundred times more lead in her body than her ancient ancestors did.

Modern history tells us the story of how powerful industry is—from cigarette companies to gun manufacturers—all over the world, more powerful than the collective voices and protests of individuals. The lead industry used its towering advantage over its victims to silence them and shame them for their lack of power. In 1955 the director in charge of "health and safety" for the Lead Industries Association described childhood lead poisoning as primarily a "slum problem" and a "major 'headache'... the only real remedy lies in educating a relatively uneducable category of parents."

Twenty years later, still blaming the victims, the industry's "health" director was making even less-veiled racist remarks. "The basic solution is to get rid of our slums, but even Uncle Sam can't seem to swing that one. Next in importance is to educate the parents, but most of the cases are in Negro and Puerto Rican families, and how does one tackle that job?"

The story of lead is like the story of tobacco in some ways, except the lead industry was never forced to pay a price for its greed, for the damage it caused, or for its attempts to ruin and discredit the great scientists who spent their professional lives fighting on behalf of its victims. Those victims were more than simply innocent—they were kids. They were babies.

Now I knew, as I browsed online, why there are so many books and journal articles about the public health history of lead—from *Lead Wars*, by Gerald Markowitz and David Rosner, to *Toxic Truth*, by Lydia Denworth. It is because the story has so many twists and turns, and so many great characters.

Based on 2015 data, the Institute for Health Metrics and Evaluation (IHME) estimates that lead exposure accounts for 494,550 deaths and the loss of 9.3 million disability-adjusted life years due to its long-term effects on health. The IHME also estimates that lead exposure accounts for 12.4 percent of the global burden of developmental intellectual disability, 2.5 percent of the global burden of heart disease, and 2.4 percent of the global burden of stroke.

That is the legacy of Charles Franklin Kettering, the American inventor, engineer, businessman, and philanthropist, a holder of 186 patents and a man of platitudes and corporate pop-optimism. His

PUBLIC HEALTH ENEMY #1

work added more lead to the environment and to children's blood than any other application of the metal. It is one of the largest environmental crimes ever. Tell me again why we're naming universities after him?

THANKS TO A HANDFUL of environmental health superstars, the government began phasing out the use of lead in paint in the 1970s—first with the passing in 1971 of the Lead-Based Paint Poisoning Prevention Act, followed in 1978 by the federal ban on consumer uses of all lead paint.

Lead was finally restricted—but not totally outlawed—in plumbing in 1986. The same year, leaded gasoline was taken off the market in the United States, and it was banned in Europe by 2000. The 2002 Earth Summit called for a worldwide ban on all leaded gasoline, with a goal of total elimination by 2006. The United Nations conducted a ten-year campaign to eliminate it from developing countries, but according to the UN environment program, in 2016 three Arab countries—Algeria, Yemen, and Iraq—still added TEL to gasoline. Ugh. And for some inexplicable reason, it is still allowed in airplane fuel.

Before industrialization, children rarely had lead in their bodies. It was due only to industrial greed and convenience that it was mined and released into the environment. Even the ancient Romans suspected it was dangerous, even deadly, but we in the modern age allowed it—we looked the other way and let convenience drive policy.

Environmental health scientists continue to demonstrate the cruel impact of lead on the bodies and brains of developing children—and how even low levels of exposure can have dramatic consequences. Each new study has played a part in keeping the antilead movement going in the face of resistance from the evil and, for a time, powerful industry lobby.

Today we have a more complete picture of the neurotoxin's evil effects. Thanks to the work of scientists and pediatricians like Clair Patterson, Herbert Needleman, Philip Landrigan, Bruce Lanphear, David Bellinger, and so many more, we know lead's potential to twist behavior, attack every system in the body, erode cognition, and possibly even warp one's DNA.

Lead exposure has been linked to almost every kind of developmental and behavioral problem, including school dropout rates and criminality. A recent study looked at six U.S. cities that could provide good crime data and blood-lead-level data going back to the 1950s. Going neighborhood by neighborhood in New Orleans, it found that a rise in blood-lead levels matched a rise in incidence of violent crime. Econometrics studies looking at worldwide crime trends also show an astonishing correlation between a "lead curve" and a "crime curve." Where consumption of leaded gasoline declines, so does violent crime. The connection between lead and crime is still speculative, but the correlations are there.

And what have we learned about how much lead the body can safely handle? Medical science has now concluded there is *no safe level of lead in the human body.*

But here's something beautiful: this graph shows a steady and significant decline in lead in kids over the last several decades. These kinds of charts give us public health folks super-serotonin highs. When people tell you that government is inept and ineffective, all you have to do is remind them of the good work that the right policies can achieve.

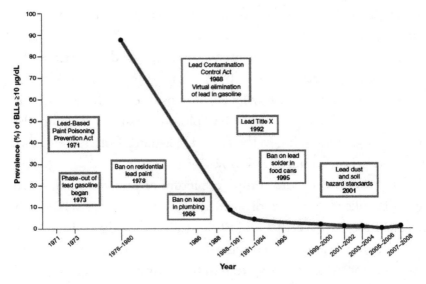

DECREASE IN ELEVATED BLOOD-LEAD LEVELS OVER TIME

Unfortunately, I must point out, the perceived "victory" over lead can create new problems. I run into older pediatricians all the time who believe the battle over lead has been won because they no longer see kids comatose from lead exposure in ICU beds. To them, the problem is fixed. Lead is a problem of yesterday. But given that we now know that even low levels of lead exposure can change our brains and bodies, our concern should be heightened and our resources increased, not reduced. In fact, given what I was learning about lead in water, the lead battle had a new front.

A PAINT CHIP, as anybody who's lived in an old house knows, comes off the wall with all its layers together. If you look at a cross-section, you can see all the layers of paint like tree rings, with the oldest paint at the bottom. And when a child who is crawling or walking finds a paint chip on the floor, it is usually put directly into their mouth, just like they put everything into their mouth.

When you open an old window, one that was painted long ago with lead paint, lead paint dust—and other small particles of lead paint—rise into the air and can be inhaled. And in summertime, the elevated outside air temperature puts more lead into the air, and more dust rises to be inhaled. Even the heat retained in the ground increases the amount of lead that is released from plumbing.

Soil is another place where lead can linger. Over the decades when lead was used in gasoline, car exhaust settled in surface soils in city schoolyards, playgrounds, and backyards, leaving them with accumulated and concentrated lead. This is why today's urban gardens on vacant land in Detroit, Flint, and elsewhere have to import their soil and often grow produce on raised beds.

These days, when we talk about lead exposure in children—in a medical setting or as a public health issue—almost always we are talking about paint and paint dust. No matter how much we hand out cleaning supplies and tell people to vacuum, mop, wash hands, or wipe the windowsills, children who have lead in their environment will be exposed to it. The problem is that lead is incredibly durable, even all these years after it was banned. It is still found in older homes. Its use was often required in public housing until 1978.

But in all my training and clinical care, both in pediatrics and in public health—and in all my work in communities disproportionately impacted by lead—I don't recall ever hearing about the possibility of lead in water.

Never. Not until the summer of 2015.

Not until the night of the barbecue, when Elin first mentioned it.

The D.C. crisis had come and gone—and since it was suitably covered up, I knew nothing about it. I didn't even know that water pipes were made of lead or were soldered together with it. I didn't know that lead is still found in most kitchen faucets, because even supposedly "lead-free" brass could contain up to an average of 8 percent lead until 2014, when regulations changed again.

To catch up on what I had missed in my environmental health education, I started reading Werner Troesken's gripping book, *The Great Lead Water Pipe Disaster*, which covers more than a century of lead-in-water disasters and the lack of political will to do anything about it. As far back as the eighteenth century, when lead pipes were installed to carry water, we've had kids with mysterious illnesses, moms with miscarriages, and adults with neurological complaints. The poisoning can affect so many systems of the body, it's hard for doctors to diagnose.

I learned that in the last 150 years, there have been numerous water contaminations—random, seemingly one-time special events in a number of U.S. cities—Chicago, New York, and D.C. In the 1890s there was a crisis in my birthplace, Sheffield, England, when the "softness" of the water caused lead contamination. Thereafter the pregnant women of Sheffield began miscarrying, which later led to the invention of a lead-based abortion pill.

In recent times, water in the United States was supposed to be protected by the Lead and Copper Rule (LCR), first issued in 1991 as part of the Safe Drinking Water Act. But as I was beginning to see, the LCR was grossly inadequate and lagged behind the curve of scientific discovery. There are an estimated six to ten million lead service lines in the country—many of them in older, low-income, minority-populated urban areas. And there are millions more fixtures with lead. The LCR did not require that these lead service lines or fixtures be systematically replaced. Studies done in the last decade have shown how crucial it is to protect developing children from any lead exposure whatsoever, but the LCR still did not deal in any way with lead exposure in daycare facilities and schools that get their water from a public water system. The action level the Safe

Drinking Water Act settled on—15 ppb—does not make any sense. It was created with water utility companies in mind—and what was economically feasible for them—rather than the best protection of children.

Why wasn't there massive public education about lead in water? Why didn't the D.C. crisis result in more public debate? And how come pregnant moms are not routinely advised to use water filters, to flush their pipes regularly, and to use only cold tap water for cooking or drinking, as warm water concentrates lead? I couldn't help but suspect that just as Kettering and Kehoe and the lead industry had won the public relations and public health war decades ago, the water utilities were winning now. Despite the well-documented history and science of lead in water, the issue remained underground and out of sight. And yet unlike any other form of lead exposure, lead in water impacts a much younger and much more developmentally vulnerable age group. Even more insidious than lead in paint, lead in water—colorless, odorless, and invisible—is meant for ingesting. In my most cynical moments, I had to wonder if the water utilities wanted the focus to remain on lead paint. Divide and distract—and conquer. It was a tactic taken straight from the Lead Industries Association's playbook.

Back in my training at Children's Hospital of Michigan in Detroit, we spent time in an outpatient clinic where kids with lead poisoning came for follow-up exams. And in the hospital itself, children were often admitted with lead levels so high that they needed full-time care and treatment. As a resident, I diagnosed and treated those kids. Kids admitted to the hospital for lead poisoning are never discharged until elaborate histories and physicals are done to determine the source of exposure.

Abdominal X-rays are ordered to detect paint chips in the gut. If they are found, they're flushed out. Nutritionists are consulted to assess dietary intake. If children who have an empty stomach or are deficient in certain nutrients are exposed to lead, their bodies will absorb more of it.

Social workers are consulted and home inspections are done,

particularly for the presence of lead paint. If it is found, the home will be cleaned, and lead will be eradicated or painted over before the child is returned home. And in cases where we couldn't find the source of lead exposure, we dug deeper. We looked at lead contamination in herbal remedies, imported makeup (kohl eyeliner), parents' hobbies or employment (stained glass, gun shops), and imported toys.

We never checked the water.

How many kids did I miss?

The eyes don't see what the mind doesn't know.

WHEN I WAS WORKING in Detroit, I cared for an African-American boy, Javonne, who had been shot with a bullet that was designed to fragment in the body in order to cause maximum damage. He was seven at the time, brought to the ER in critical condition, and he underwent extensive surgery, including a colostomy. The surgeons tried to remove all the bullet pieces, but it was impossible to get them all.

When I met him, several years later, Javonne was thin and lethargic. He was ten—uninterested in eating and doing poorly in school. His lead levels were still climbing. The few remaining bullet fragments were continuing to release lead into his body. So he was sent back to surgery, and more fragments were found and removed.

My memory of Javonne stayed strong in my mind. Years passed, and I still remembered his stoicism and hesitant smile. I thought about everything he'd been through, largely preventable, all the toxic stresses: violence, fear, bullet wounds, hospital visits, surgeries, and PTSD, and then the effects of lead poisoning. For many people, life isn't long enough to recover from a childhood like that.

What Field Are You On?

ELIN MAY HAVE BEEN VEXED ABOUT MARC EDWARDS AND HIS reputation for trouble, but for the last two days, since his water findings were released on September 8, my gut had been telling me that I should get in touch with him. He was a scientist who had worked on the front lines of the worst lead crisis in history—one that had been covered up by authorities for years. He knew what being stonewalled was like. We couldn't let that happen in Flint.

But every time I mentioned reaching out, Elin became anxious. She kept saying it was risky. I kept thinking, *My God, he's a scientist, not Hannibal Lecter. How dangerous could he be?* On the morning of September 10, after Jenny and I reviewed the results showing a rise in lead levels, and Brad Wurfel was all over *MLive* attacking the Virginia Tech team and calling them irresponsible, I nagged Elin into submission. The Hurley clinic's blood-lead results were way too sensitive to share with anybody—I was too nervous to even email the results to Elin—but I felt emboldened and confident. We were on to something big. It was time to start laying new groundwork.

Elin emailed Edwards and his team, using the addresses listed on his Flint Water Study website. She thought it was much better if she played go-between, so if things became dicey, and Edwards became difficult, I would not be implicated. Besides, Elin had her own reasons to be in touch with him: she was an alum of Virginia Tech, an environmental engineer—and a water expert herself. The fact that she had never met Edwards personally didn't matter. She knew of him. Everyone in the water world did.

That morning she wrote to Edwards and his assistant, a graduate student named Siddhartha Roy, saying that she had a friend, a doctor, who was looking at blood-lead levels in Flint and was designing a study.

Edwards replied right away:

FROM: Marc Edwards
TO: Elin Betanzo; Siddhartha Roy
SENT: Thursday, September 10, 2015, 9:44 A.M.
SUBJECT: RE: Flint—Blood Lead Level data

Hi Elin,

Excellent. The information I think she should know is that there are three time periods of interest. Obviously, before the switch to Flint water. Then, after the Flint River water was introduced, lead did not rise immediately. While we cannot be sure exactly how fast it rose and when, it was starting to become a problem by late 2014 for sure. But it became a much bigger problem by February 2015 to the present. As a rough linear estimate of the risk for each time period based on the data I reviewed, I would guess that Detroit water is a 5, Flint from the switch to late 2014 is a 10, and Flint from early 2015 to present is a 25.

The other thing is that unlike lead paint, the risk from lead in water is highest in children under age 15 months, and is especially high for infants drinking formula. The normal blood-lead monitoring program is concentrated on children with the highest lead paint risks. So to the extent she has data for the at-risk children under age 15 months, they should focus on that if they want to find the true health impacts.

Marc

I was impressed. He was worried about the littlest kids. In the afternoon, when my name was included on an email exchange that Elin had started with Sid and Edwards, I couldn't stop myself and just jumped in.

FROM: Mona Hanna-Attisha
TO: Elin Betanzo; Sid Roy; Marc Edwards
SENT: Thursday, September 10, 2015, 2:56 P.M.
SUBJECT: RE: Flint—Blood Lead Level data

Thank you all for the below info! We are currently analyzing the blood-lead levels of our patients. We routinely do blood-lead levels at both the one-year and two-year well child visits for all kids. We see the largest number of Medicaid kids in the county—and yes we are focusing on the infants in the attached zip codes and comparing those levels from the pre-switch period to the most current period. We hope to have some analysis back in the next day or so. FYI we still have quite a few kids with chipping paint issues and retained bullets that cause elevated leads.

I also explained to him our difficulty in getting the larger sample from the county and state and closed with:

Mr. Edwards, I understand that you will be in Flint next week. Let me know if you have any free time—would love to meet and discuss more.

    Thanks!

Elin was horrified and back-channeled me immediately: *What are you doing?* But in less than five minutes, Edwards replied.

FROM: Marc Edwards

TO: Mona Hanna-Attisha; Elin Betanzo; Sid Roy

SENT: Thursday, September 10, 2015, 3:00 P.M.

SUBJECT: RE: Flint—Blood Lead Level data

Great. Two weeks ago I submitted a formal research request on blood-lead records, to the health department, for the records in Detroit, Flint, and Genesee County. If I get it will let you know.

Yes, Sid and I will be coming up, and we have some availability on Tuesday afternoon. Is there a time that works for you? Maybe we could do lunch?

Marc

So Edwards was going after the same data—and hadn't gotten anywhere either. And he accepted my invite. A lunch date! My heart was pounding as I responded. We quickly settled on a place and time to meet. Elin continued back-channeling me with warnings. But I needed to meet this guy, see him face-to-face, to figure out whether he was trustworthy and legit. Whenever I pictured him, in my mind he was wearing the panda necktie. I couldn't help but smile. He must be brave to wear something like that.

JENNY AND I WERE talking every hour and emailing and texting throughout the day about the research we were trying to design. Every hour or two, I'd have a new thought about the study—and how to make it as accurate as possible. Impatient, I would jump out of our text exchanges and, unable to type as quickly as I could talk, just call her. "Jenny, I'm so sorry to bother you, but . . ."

The study had to be perfect, otherwise it would be tossed out and dismissed or else cause unnecessary alarm in a city that was facing too many challenges already. We kept improvising our way through trial and error to get it as tight as possible, based on every idea and piece of information that came our way—there were new developments and twists every few hours. Jenny and I talked between meetings, while composing other emails, while driving, while walking, while in the grocery store, and while she was breastfeeding. When we couldn't speak, we texted. We made plans to work over the weekend, getting the IRB application ready and preparing to run the data. It was a short period of time to design such an important study, but we didn't have a choice. Every day mattered.

Of course, the rest of life didn't vanish while I obsessed over our study. The stiffness and pain had returned to Elliott's right shoulder and were possibly even worse. He was still unable to sleep, drive, or return to work. He'd even given up trying to lift the heavy lid of his beloved Kamado grill. And when the girls had soccer practice, the job fell to me—which was the case on this evening.

I'd missed a few games already—and my absence was noticed. This was a dramatic change from the previous year, when I was the assistant coach for Nina's team, a job that I was qualified for because for a couple of years in high school I'd been a goalie (and a terrible one) and because I knew the girls' names. This year, thankfully, another dad stepped in who truly knew the sport.

The practice was at a local high school—it was practically in Elin's backyard. So before leaving for the field, I printed out the clinic blood results, thinking that I'd swing by her home and show them to Elin in person. It seemed too sensitive to email. I hadn't even wanted to go over the results on the phone. I stood by the printer and waited for the pages to come out, then put them inside an envelope and dropped it into my bag.

I did my best to track down and gather up all the assorted soccer gear the girls would need. (In a future lifetime, I'd like to invent an all-in-one soccer cleat, shin guard, and sock combo that also functions as a water bottle.) Elin texted that she was going to meet me at the high school field instead and bring her eight-year-old son, Vincent, who had a crush on one of the girls on Layla's team.

As soon as I arrived at the high school, there was another text waiting from Elin, a sweet picture of her daughter Gabriella standing by the front door in a pink ballet leotard and tutu. It was one of those mom-to-mom moments, old friend to old friend, that grounded me. I've known Gabriella since she was born. I am one of her godmothers, in fact. Seeing her smiley young face was just the tonic I needed.

ELIN: What field are you on?

ME: 3

ELIN: Coming

As the kids were scrambling onto the grass, learning how to pay attention and kick the ball, and the other moms on the sidelines were talking about piano lessons, school recess dramas, and homework struggles, Elin and I stood at the intersection of many fields with folding soccer goals set up. Hundreds of little kids were running around, kicking balls, cheering.

I looked around nervously, as though that awful Brad Wurfel might suddenly appear at any moment and hit us with a demoralizing quip.

Then I pulled the printout from my bag and handed it to Elin.

I watched her closely as she read—and waited for her reaction.

MOST OF MY PROFESSIONAL LIFE is pretty predictable—to me, at least. Clinics, meetings, presentations, red tape, trainings, and lots of talking with faculty and residents and patients. It's the life of a doctor, the life of a healthcare professional, the life of an educator. But now it was becoming something else—the life of a renegade and detective. And standing there on Field 3, with our backs to the kids, huddled like two spies, my life was suddenly beginning to feel like an episode of *Scandal.*

I'm not much of a TV person. I don't have time—and we don't even have cable at home. But Elliott and I had just binge-watched all the past episodes of *Scandal,* getting ready for the new season that was supposed to start later that month. I love the main character, the fixer, Olivia Pope, and how strong, smart, and beautiful she is. And I love how, in each episode, problems arise and go deeper and darker and more wide-reaching than first imagined. But then, in every episode, a solution is found. Just as in an episode of *Scandal,* the deeper I dug into the Flint water issue, the darker it got. I was hoping the scriptwriters had a bang-up solution coming.

Elin finally looked up from the printout—and grabbed me.

She was too stunned to speak.

Back at Kimball High in the environmental club, we had wanted so passionately to save the planet, to right wrongs, to make a difference. That seemed so long ago. As the years passed, in our own quiet and methodical ways, step by step, we pointed ourselves in that direction, always tried to do good work and to make a difference. We stayed activists in college, and in our careers, but accomplishments in the adult world turned out to be harder to achieve and far less exciting than our wide-eyed, swashbuckling teenage activism. The problems we found now were mired in complexity, and victories were faded by trade-offs. So far very little of it had felt as good as keeping the incinerator in Madison Heights shut down, organizing a Greenpeace conference, or protesting a nuclear storage facility adjacent to the Great Lakes.

But in that moment, as we stood on the sidelines of Field 3 and looked at the clinic results, I think we both realized that all our calls, emails, texts, and strategizing of the past two weeks had been for something. Something important.

Elin is usually pretty reserved, not a screamer or crier, but her reaction now—her hand squeezing my arm tightly—was as demonstrative as I'd seen her since those old days.

"Numbers don't lie," I said.

"No, they don't."

"Flint kids are being poisoned." All my rage and sadness returning, I became so upset that tears welled up in my eyes.

Elin nodded. I could see she was angry too, enough to cry with me. It wasn't just about Flint. It was also about D.C. and all the things she'd seen there, the negligence and butt-covering, the wrongful actions by people in positions of public trust. She had been living with what she knew—and maybe even with a sense of guilt for having been at the EPA while the crisis was unfolding but not doing more about it.

"What are you going to do?" she asked me.

"I don't know," I said, thinking for a second. "I can't share this—not with anyone. Not yet. I need more data, a bigger sample. I'm working on that. I don't want to get it wrong and blow my chance."

Patience isn't my strength, but I needed it now most of all. I wanted to release the results as soon as possible and get the state to deal with this crisis, but that would require a framework of support and unassailable political backing. I couldn't go it alone. I needed a team.

"Don't let Edwards see anything," Elin said.

"Don't worry."

"Jordan in Representative Kildee's office should be the first, if anybody."

"I can't share anything yet."

Elin considered that, just as a wayward soccer ball sailed toward us and we silently dodged it, still lost in thought. My younger daughter, Layla, looked over, laughing, and ran away.

"The whole thing is so hard to believe," I said, shaking my head. "What is wrong with MDEQ? Isn't safe drinking water their job?"

"They must be idiots," Elin said. "Clear disregard for people—and public health."

As soon as I got home from soccer practice—cleats and shin guards off, the girls showered and fed—I sent an email to Melany

Gavulic, the Hurley CEO and my boss. I had written her a couple of days before, on September 8, to let her know about the investigation and that our first data set from our clinic patients showed an increase in lead levels. Now I needed to follow up.

Melany is a Flint native, a Kettering University graduate, and a former nurse who rose steadily in the ranks at Hurley to become CEO in 2012. She's a strong, levelheaded, and insightful leader, and of course, I hoped she would support what we were trying to do. So far, I had had no indication that she wouldn't. But Hurley is a city hospital, and our board is political—each member was appointed by the city. As CEO of a public hospital, Melany had to placate an array of conflicting agendas and a lot of strong personalities who were city movers and shakers. My findings had to be carefully and skillfully handled, because there was a lot at risk, not just to my job, reputation, and credibility, but to the reputation of Hurley itself.

Looking ahead, if the IRB application was approved and I received permission to run the county numbers that were stored at Hurley, and if Kildee's office helped me obtain access to data from the state, what should I do next? Considering the deafening silence coming from the state and county health departments, and the attacks on Edwards—and all the previous attacks on residents who tried to raise concerns about the water—it was painfully obvious that this was going to be a political issue. But how political, and how heated?

I tried to guess at the timeline. It was Thursday night. Jenny and I would be writing our application to the IRB over the weekend, then we'd submit the data request. Once we got the data, we'd review and crunch the numbers, and then we'd probably want to double- and triple-check our analysis before we shared anything with anyone. Jenny had already heard from the folks in the IRB office that they were determined to move the process along as fast as they could. Rather than spend one or two weeks to review our application, they might take just a few days. After that, I suspected everything would have to be cleared by Hurley's legal office, and maybe the PR folks would get involved in some way.

Things could start happening quickly after that.

"I think all this is going to significantly explode next week," I wrote to Melany. "We can be part of the chorus (for health and kids) or an absent voice."

Within six minutes, she wrote back.

"Oy! This is a political mess and given our relationship to the city . . . we could really create a mess if we don't give them a heads up. Alongside that, we are the children's hospital in this region and we have a duty to advocate for child health. Do you have some time tomorrow to talk a bit more about this and what you know will be coming out? I'd like to pull in some others and certainly not blindside the city."

PER HER REQUEST, JENNY and I had an urgent meeting with Melany the next day, Friday, September 11. We prepared a short summary of our preliminary clinic results and brought Melany up to date on the process so far—namely, how many ways I had tried to get more data from the county and state, and how many dead ends I'd encountered despite my continued requests. I was still trying to figure out exactly why I kept running into dead ends, and every theory I had was alarming.

The health departments were blatantly covering up, stalling, or inefficient and slow to respond, or they truly didn't care. It could be laziness or, worse, an indifference to poor black and brown people. They didn't hear the sirens I was hearing, while I couldn't get them out of my head.

Melany worried about the study being leaked or released before we were ready. I told her about the lunch meeting with Marc Edwards scheduled for the following Tuesday, and I assured her I would not share the clinic results with him. I knew we needed to control the release of information and make sure our findings were unassailable.

Afterward I went back to my office and started pacing. There was so much going on, so much to think and worry about. I hadn't slept well in days—two weeks now—and I was having weird dreams about lead; my mind never stopped thinking about the study or strategies. The knot in my stomach wouldn't go away. I was on a coffee-only diet. Some people might stop drinking so much coffee as a way to calm down. But the only thing that worked for me was moving forward—doing something, trying a new angle, pushing every button.

So I stopped pacing and called somebody I'd been thinking about lately, an older superrespected pediatrician and friend. Dr. Lawrence Reynolds was the CEO of Mott Children's Health Center, a nearby facility where lots of Flint kids are also treated. An African American who has always been a champion for underserved kids, for reducing racial disparities, and for pediatric advocacy, Dr. Reynolds had served on so many state boards, it was hard to keep track of them. They included a childhood lead-poisoning commission and an infant mortality initiative, and he was just wrapping up a stint as president of the Michigan Chapter of the American Academy of Pediatrics (MIAAP).

If I wanted to get anywhere in Flint and with city politics—which I didn't really know—I would need his support and his gravitas in my corner. Whatever our study showed, and whenever it was released, having Dr. Reynolds behind it would give it more credibility and impact.

He answered the phone, but I could barely hear him. His voice was foggy and thick. He explained that he was sick—battling a bad sinus infection for the last week. I apologized for bothering him, he was clearly quite ill, but I kept talking. "Dr. Reynolds, have you heard about the water? There are news reports of lead—by a Virginia Tech researcher—and my friend, a water expert, has shared some information with me about this." I wasn't sure how far to go with it.

"No, I've been sick all week. I haven't heard a thing."

"Well, let me ask you this. Any chance you guys have looked at your lead levels in clinic? We are looking at our levels."

Lead. At just the mention of it, his voice grew stronger, almost as if he were no longer sick. He asked me what I'd learned.

I couldn't reveal that we'd pulled up results already, because he might ask what they showed. "I'm not having a successful time getting blood-lead data from the county or state," I added, and shared my frustration about the jurisdictional breakdown—how the presence of lead in water somehow made it a public works issue, not a concern of the health department.

We hung up quickly after that, and he got to work right away. Before an hour had passed, he sent me an update. I had a comrade-in-arms.

FROM: Lawrence Reynolds
TO: Mona Hanna-Attisha
SENT: Friday, September 11, 2015, 1:14 P.M.
SUBJECT: Re: Lead levels

Will have to check for HIPAA compliance and some advice from your research staff, clarifying the use of the data. In the meantime we can do our own query for the same data base.

I left messages with the state. All out until Monday.

Working on DEQ water quality people next.

EVERY FRIDAY AFTERNOON AT Hurley, we have something called "resident conferences," when medical experts in different fields cover a topic. That day's conferences were about pediatric gastrointestinal issues. My office is right next door to the classroom, and as usual, I slipped into the conferences and sat up front. It was my chance to see everybody each week and to be seen. I liked being a fixture at the conferences and prided myself on being intimately involved in my residents' training.

But that week, as I listened to the reflux discussion, I felt the knot in my stomach tighten. I couldn't remember the last time I'd had a real meal—or eaten more than a few bites. I was losing weight. My pants were baggy and my white lab coat was starting to swim on me, as if I'd borrowed it from somebody else.

During a break between lectures, I slipped out to make a call in my office and was followed in by Allison, my second-year resident whom I'd known since she was a second-year medical student. I've always had an open-door policy, trying to be accessible to my trainees.

I could tell from Allison's face that she was worried about something.

She closed the office door behind her. "Are you okay, Dr. Hanna? Is there something wrong?"

That threw me off guard. This was the kind of question that I usually asked my residents, not the other way around. "Just busy with the water thing—the research."

She nodded sympathetically, but wasn't convinced.

I sighed and attempted a reassuring smile back.

She was going to be a great doctor someday—I could tell that much. A Flint native, she could have trained anywhere for her pediatric residency, but chose to stay and serve the city where she was raised. She is skilled and smart, a hard worker, and has all the science down. But more than that, Allison has incredible emotional intelligence. She is fantastically sensitive and empathetic. She was the first one to notice when another resident was struggling—and the first to alert me. Now she was sending out a new kind of alert. And it was about me.

—

CHAPTER 13

The Man in
the Panda Tie

O N TUESDAY MORNING, I STOPPED BY JENNY'S OFFICE TO SEE if she'd heard back from the IRB about our application. On her phone, she was watching Marc Edwards in front of Flint City Hall. Standing in a double-breasted suit, he was holding up two bottles of water: one was labeled "Flint" in Sharpie marker written on jagged tape, and the other "Detroit." It was meant to show the difference in clarity. He explained that Flint's water was nineteen times as corrosive as Detroit's. The iron pipes in Flint were corroding—and turning the water rusty brown.

"Are you sure you don't want to come to the lunch today?" I asked her.

"I'm sure."

"I know you want to meet him. I know you're curious."

We'd been investigating Edwards together, forwarding articles and sending links back and forth. We were both intrigued.

"I want to go, but I'll share too much," she said, her face turning red. "I can't be trusted. You go. And tell me all about it."

We'd worked all weekend and Monday nonstop, like two symbiotic organisms, to finish our research proposal, complete the necessary patient protection forms, and get our IRB request ready to submit. It was one of those mind-tunneling experiences of total focus, like being in an isolation tank or studying for a big math test. On Sunday morning, when I confessed to Jenny over the phone that I was having bizarre dreams, her voice jumped.

"About lead? Oh my God, I've been having them too!"

MY MOM WAS BACK from D.C., so at least we had Bebe—along with her craft projects, her hand-rolled dolmas, and the sweetness and love she brings wherever she goes. Within minutes of returning, even while complaining about how messy we were, she had transformed our house back to an organized, pristine place, and magically, the freezer and refrigerator were stocked with food.

But not even the fragrant stew on the stove—*fasulia wa timen*, green beans and onions in a tomato sauce, one of my favorite Iraqi dishes—aroused my appetite. I burrowed into my bedroom, out of sight, to work with Jenny by phone or by email. Every so often, I'd look up and Nina would be standing at the doorway, quietly, kind of forlornly, with her big, dark eyes.

Or I'd hear Layla coming down the hallway, hollering. She had a totally different approach—always out there with her feelings and demands, always the first to speak out.

"Mama, come on—Bebe's making *fasulia wa timen*, your favorite."

"Mama, is your computer more important than me?"

"Mama, I miss you."

When Jenny had to feed Drew, or herself, or take breaks to pump, I tried to engage with my own family. I wandered downstairs and found them watching TV or playing a game. I gave them hugs and smiles, tried to be there. But my mind just wasn't. My stubborn, goal-obsessed self had taken over. I needed to protect my Flint kids, and to do so, I had to create the most perfect, most unassailable study about children's blood-lead levels that I could, one that nobody could shoot down.

BEBE WAS THE SECOND PERSON, after Elliott, to notice how skinny I was. I'd given up on most of my pants by then and was wearing sweats with a tie waist or elastic, so they wouldn't fall down. After being in D.C. for a week, the first thing out of Bebe's mouth was "You lost weight?"

"I have?" I said, trying to fake it. I'd been skinny my whole life, but after having the girls, some of that residual baby weight had stuck around. Years passed, and I kind of stopped imagining being thinner. To be honest, I was kind of happy about the weight loss.

Bebe cocked her head and studied me a little closer. "What's going on?"

"Nothing." I shrugged.

"Men saduk?" Bebe wanted the truth.

"The usual stuff. September is always busy. *You know.*"

I wanted to explain what was happening, but I knew I had to take special care with Bebe. She falls easily into dark holes of anxiety and dread. Anything I said would spur a race of new worries in her mind.

The ghosts of Iraq were hard to beat down. She would fear that my life was in danger. She might worry that I could "disappear," the way some of her friends and family members had in Baghdad, especially those from disfavored groups, like Shiites and Kurds, and leftists like my parents. Bebe had seen too many bad things, or heard about them, to trust authority. She'd seen political violence and sudden death. I remembered going to a dissidents' meeting with my parents when I was young and noticing an old man who was walking with a pronounced and visibly painful limp. Bebe told me later that he was a poet and had been tortured, electrocuted, for writing poetry.

Experience had taught her that leaders can be evil and do evil—not just Iraqi leaders but people in power anywhere, anyone who sees people as disposable instruments in their own plans.

I had more faith and, I guess, more trust. But a part of me too knew what Bebe knew. Mark and I grew up surrounded by stories of her sorrows and losses. We felt them, absorbed them, as if they were our own. I had no illusions about humankind and its potential for carelessness, selfishness, and misuse of power. My parents had taught us, without even needing to say the words, to believe in a better future and to work to create it—one that was fairer and kinder, that drew from the best in us.

But first she would worry.

ELIN ARRIVED AT THE Hurley clinic in the late morning. It was her first time in Flint—I was surprised to learn she hadn't been to AutoWorld with her GM family. I gave her a short tour of our new clinic and introduced her to a few residents. Then we braced ourselves for lunch. We were meeting Marc Edwards and were both a little nervous.

Lunch would be at Steady Eddy's, my favorite go-to place inside the Flint Farmers' Market, just downstairs from the clinic. Located next to the stalls of fresh produce, local honey, and a great deli, Steady Eddy's is one of those classic crunchy hippie cafés with sprouts in the sandwiches, vegetarian chili, tofu, and fantastic, fresh multigrain bread.

Elin and I arrived first, followed by Dean Dean. I had invited him too, wanting to bring in his background, experience, and military commando confidence. I had no idea that a decade earlier, he'd had several tense encounters with Edwards, in Dean Dean's days as the Ingham County health officer in Lansing. Elin shook his hand, and I covertly looked around, then pulled my clinic results from my bag to share with him. He read the pages carefully, nodding. His brow furrowed as he asked about the results, pre– and post–water switch.

"What was the standard deviation?" he asked.

I saw Marc Edwards making his way toward us and quickly put the results back in my bag. He was gaunt, wearing the same oversize suit he had been wearing in the video I saw on Jenny's phone and the same necktie with Serengeti water buffalos on it. He was with Siddhartha Roy, his skinny PhD student. They both looked tired and rumpled, as if they'd just been through something awful. Sid was lugging a video camera and tripod—not something you expect an engineering PhD student to keep in his toolbox—which I guessed he'd needed at the Flint City Hall press conference.

We said our greetings, moved the chairs around, and nervously reached for our water glasses. But then we stopped. Nobody wanted to sip the water.

Dean Dean studied Edwards, maybe waiting for a moment of recognition, but there wasn't one. In 2004 there had been reports that Lansing had elevated lead levels in its water. At the time, the conventional wisdom propagated by the CDC was that lead in water was a minor concern—very minor compared to lead paint exposure. However, Edwards argued otherwise and was brought in by state senator Virg Bernero, who later became the mayor of Lansing. Dean Dean didn't buy Edwards's analysis at the time.

Edwards and Dean Dean had had heated exchanges about Lansing's lead service lines, and publicly traded letters to the editor. But ultimately Virg Bernero and Edwards won the battle and put in a plan to get Lansing's lead pipes replaced. It took a decade, but Lansing is now the only city in the country besides Madison to have replaced all its lead service lines. And that was only sixty miles from Flint. In 2010 Virg Bernero ran against Snyder for governor and obviously lost. But what if he hadn't? I couldn't help but think that the Flint water switch never would have happened.

"This is Dean Sienko, from Michigan State University's public health program," I said to Edwards and Sid. Edwards leaned in and nodded blankly, not making the connection to the battles in Lansing. He was nice enough, pleasant but a little awkward, perhaps uncomfortable or just untrusting. He was focused on Flint now anyway. He was pessimistic when describing the press conference and how little he felt he'd accomplished.

When I mentioned fixing things and making a difference and trying to get everyone positive and motivated, his fixed stare cried out, *Get real!* or *What sunshine have you been drinking?* "This kind of thing takes a long time," he said more than once while shaking his head. "I'd be surprised if anything we do makes a difference." He expressed such a strong distrust of government—the state, the federal players like the EPA and the CDC—that it shocked me. He seemed bitter and wounded—and angry.

I could feel the weight of his baggage almost instantly. The doctor in me wanted to heal him, to calm him, to take the pain away. If only I had something in my doctor's bag that could help.

Dean Dean, on the other hand, sat quietly, pleasantly, and just listened. I tried to guess what he was thinking but couldn't. We ordered lunch and I got my usual, a veggie melt on pumpernickel, which I doused with the Cholula Hot Sauce that was on the table. I was a vegetarian in college and throughout medical school—not really for health reasons; I just didn't want to eat anything that had to be killed. But eventually I had caved and consumed the occasional piece of chicken or seafood to please Elliott's mom, Mama Jeeba, a prodigious cook who puts meat in basically every dish she makes. It felt rude, and kind of selfish, not to.

The restaurant was busy, full of noise and distraction. We all leaned forward to hear one another, so we wouldn't have to talk so loudly that we could be overheard. Edwards said the lack of MDEQ oversight violated both the Safe Drinking Water Act and the Lead and Copper Rule. He was pretty sure of that. He explained that after the D.C. crisis in the early 2000s, new regulations had been put into place, ensuring that anytime a water source is switched, significant sampling needs to be done before and after. This didn't happen in Flint. Instead, the agency was gaming the system and sampling in a way to get the results it wanted. It was unclear why federal rules had not been followed, he said, but they should have been.

"Why isn't the EPA taking over from MDEQ?" I asked Edwards.

"The way the law is written, MDEQ is in charge," he explained. "It's called primacy."

I half-remembered that word, *primacy,* from my environmental policy classes at the University of Michigan. It had to do with how much control states were given in enforcing federal environmental protection laws. It had come up in recent conversations with Elin, but with so many other things to learn about water, I hadn't asked for clarification.

"The EPA gives state environmental departments a pretty wide latitude," Edwards continued, "and often defers to them, even when it shouldn't. I would argue, given all we know, that the EPA has the authority to take over now from MDEQ, but it probably won't."

"Why not?" I studied his face for signs of where he would go next, but he was hard to read.

"My experience is," he replied, "the EPA and the states work hand in hand to bury problems. EPA rarely challenges the states."

I had feared that but hated hearing it.

Edwards went on to explain the water science to us, easily going from small but important details to the big picture. The Flint River's water was innately corrosive, he said, which presented a lot of challenges that the city didn't have the capacity to handle, in terms of personnel or expertise. He knew of a similarly corrosive water supply in Maryland, where even a staff of thirty water engineers couldn't manage to control it properly. Compare that with Flint, where only one half-time person had worked on the city water prior to the switch. Supposedly more people were working at the treatment plant now, but still not enough.

The water from Detroit, which is Great Lakes water, was less corrosive, but even it was treated with corrosion control. Edwards felt that until Flint was restored to treated Detroit water, its old source, its tap water would remain dangerous.

Plates of food came, and nobody ate. I suspected I wasn't the only one with a big knot in my stomach, and my knot got tighter every time Edwards spoke. The water in our glasses still sat undisturbed. It was too sad to even make a joke about.

We leaned in closer and tighter, trying to hear and be heard. I think we were trying to decide if we trusted one another. I had to marvel at how different Edwards and I were, basically opposites. He is a tall, older white guy who looks like his ancestors came over on the *Mayflower*. I'm short and brown, an immigrant from a country that the United States has basically been at war with for three decades. He resided in a place that was in the heart of the Confederacy. I lived so far north, it was practically Canada. He talked skeptically of government, while I believe in government's capacity to help people—and even to remedy historic wrongs and injustices.

But we shared so many things too—a fierce love of good government, transparency, ethics, and integrity in science, and most important of all, an abiding, almost debilitating concern for children. We were allies. The way I cared about our kids in Flint, he cared about kids everywhere. And he knew they were in danger. He could see colossal lead-in-water problems in cities all over the country. And he could see that they weren't being addressed.

Trust. For me, it took only a few minutes to develop. Edwards's philosophy was not just about righting wrongs—it was broader and bigger. He was a deeply moral person. And he wouldn't be a bystander to injustice. He was in Flint because he took it as his duty. Solving a public health crisis should have been the business of the people we'd empowered and entrusted and funded with our tax dollars to keep our air and food and water supplies safe—and not left to an out-of-state engineering professor or a pediatric residency director. But for Edwards, it wasn't a choice.

He and Elin started talking about water engineering stuff. It was water geek jargon, almost incomprehensible to me. I understand medicine and science, but the technical engineering lingo is slightly out of my comfort zone. *Percentage of water leaks, types of corrosion inhibitors, disinfectant by-products . . .*

I kept peering into my bag, looking down at the copy of my clinic results, the preliminary numbers. I wanted so badly to show him—or tell him. I thought about Jenny—would she have been able to resist this moment? I looked at Elin, who knew what I had in my bag. As did Dean Dean.

But rather than say anything, I asked a hypothetical: "What if someone looked at the blood-lead levels *and* saw an increase after the water switch? Would that make a difference?"

Without a pause, he spoke with total confidence. "That would be a game changer."

Elin and Dean Dean and I shared a look.

I tried to catch my breath, while Marc—because he was Marc to me now, not Professor Edwards—began talking again about how the most vulnerable children are babies on formula, because it is most often mixed with tap water.

The proper ratio of powder to water is one scoop for every two ounces of water. But some new moms found it confusing and diluted the formula with too much water, sometimes on purpose, hoping to make it last a little longer. Watery formula can result in poor growth, low sodium, and sometimes even seizures. Now, on top of those concerns, in Flint at least, adding more water also increased an infant's exposure to lead.

There were ready-to-feed formula options, something I mentioned to Marc, who seemed very interested and asked for more information. He was an incredible listener, always leaning in closer. I explained that the ready-to-feed formula came in bottles and was already mixed—no water needed—but it was also more expensive and not usually covered by federal food programs for the poor.

This got us brainstorming. Was there a way to get the ready-to-feed formula to Flint babies? We talked about getting a waiver from the USDA to give out ready-to-feed formula at the clinic. Elin mentioned that she might try Representative Kildee's office for help with that.

As the lunch was coming to an end, and totally out of the blue, in sort of a non sequitur, Marc blurted out the words "I trust you" to me. He hadn't been sure he could when we first set up the meeting. He assumed we were cut from the same cloth as the other "public health folks" who'd undermined his work and kept lead in water for years. But to his surprise, he'd come to Steady Eddy's and found allies.

And I had a new friend.

I RAN BACK UP TO CLINIC—sometimes I still can't get over the fact that I work above a farmers' market—and quickly sent an email to the folks who run the Genesee County Health Department's Women, Infants, and Children (WIC) program, which provides pregnant moms and young kids with nutritional support. Could they help us get ready-to-feed formula? I thought of little Nakala, now on her fourth week of powdered formula. The clock was ticking, not just for her but for all the babies in Flint. If there was ever a time I wished we had more breastfeeding moms, it was now.

"Babies who are drinking formula mixed with tap water (especially warm tap water) are at greatest risk for lead exposure," I wrote. "Water is such a huge proportion of their diet and their brains are still developing. Until we have definitive information about lead exposure, we should err on additional prevention, especially for our most vulnerable kids."

I received a response that they were concerned, but "don't have the means to provide either water or vouchers." Angry, I texted Elin to see if she could follow up with Jordan in Kildee's office, hoping he could pressure the USDA to provide ready-to-feed formula now. She did, and Jordan tried but was told that the USDA couldn't do a waiver unless there was actually a health advisory or official emergency.

So that was it: we were back in the Bizarro World of the Kehoe Rule, where we had to prove there was danger and harm before anything could be done. The onus was on us.

I summarized my lunch meeting in a long email to Melany. Thanks to all my talks with Elin and now Marc, I felt like I had taken a crash graduate school course on lead and water. "We're facing a disaster," I wrote. "It could be seen a mile away, but now that it's in our faces, we have to act, and soon."

Melany responded quickly. She thanked me for the recap and said she was already thinking about the next steps. Then she said the magic words.

"I want us to be part of the solution."

THAT NIGHT AT HOME, Elliott was happy to hear how the meeting had gone—and how Melany had responded. After he fell asleep in the recliner in the corner, I opened my laptop. It was still September 15, after all. Even without Marc Edwards, the meetings, the emails, and the submission of the IRB application, it was one of the biggest days of the year for me.

It was the day ERAS opened.

ERAS is the Electronic Residency Application Service, which transmits medical student applications to prospective residency programs. So on September 15 we learn who has applied to become a resident at Hurley Medical Center. The fruit of all my work during the year ripens—and I can see who is applying to our program. Every year we want more and better applications. In recent years, since we had been working hard to upgrade the program, we were up to thirteen hundred or so applications per year. That number is whittled down to a list of two hundred applicants whom we invite for interviews to fill our seven slots.

I'm the one who does the whittling. As soon as ERAS opens, I start searching for the best applicants. In order to get a running start, I try to read one hundred applications a day between September 15 and the end of the month.

When I came to Hurley, the pediatric residency program was falling apart. Some unhappy residents were transferring out. Now with a revamped curriculum and several key hires—mostly women faculty—our morale was fantastic. Our residency theme song was taken from *The Lego Movie*: "Everything Is Awesome." We were attracting a more competitive and diverse group of residents, ones who increasingly looked like our patients. They came from all over—caring, idealistic, and passionate new docs who wanted to change the world. And they wanted to live and work in Flint, which shows how dedicated they were. Lying in bed, I began reviewing applications, but it was hard to turn any of them down. I loved them all. How could I not?

With my laptop open on my bed, I quickly fell asleep.

THE NEXT DAY OUR IRB application was approved. We had submitted it on Tuesday, and it was approved on Wednesday, an unheard-of turnaround time. Hurley may not have dozens of NIH-funded researchers and an entire building dedicated solely to research, but we don't have the sputtering and stalling bureaucracy that often comes with all that either. We are small and lean and responsive. Everyone knows everyone. And the credibility that Jenny and I had built over the years carried weight.

"Mona," Jenny teased me, "has anybody ever said no to you?" After two weeks of hearing "no" from everyone I spoke to, I could only laugh.

Environmental Injustice

WITH THE IRB APPROVAL LETTER IN HAND, JENNY AND I NOW could access the blood-lead levels obtained by other clinics and doctors in Flint—at least all the labs processed at Hurley.

In a matter of minutes after we submitted our data request, at exactly 7:53 A.M. on September 17, 2015, our IT department sent us a full data set pulled from the Epic electronic medical records. We went from a sample size of about 350 to one of almost 2,000. Before EMRs, this kind of study would have taken months, possibly years. It would have been painstaking work, reviewing paper chart by paper chart by hand. An analysis still took time now, but getting the data itself took only minutes.

I didn't know what the exact results would be, but waiting wasn't an option. About ten minutes after looking over the new data set, without permission or even any hesitation, I blasted out an email advisory to my residents and faculty at Hurley. My biggest concern was the babies, who were the most vulnerable. I sent out a set of instructions for the clinic to provide to all the moms, dads, and other caretakers:

All,

There has been a lot of news recently about the possibility of
lead in the City of Flint water. . . .

Until more information is garnered, I think it would be
best to take some precautions for our most vulnerable patients.
Babies on formula consume a huge amount of water. So at
your clinic visits and in the newborn nursery, please continue
to recommend 1) breastfeeding! and 2) if using formula, do
NOT mix with tap water and especially not with warm tap
water.

This was the best I could do for now. I knew I had to do *some-thing*, even though no one at the state or county health department
seemed particularly interested in the subject of lead in the water, or
had done much to take action. Never mind that I was a lifelong
progressive and believer in government—and was quickly losing
faith. A sea of red tape lay between me and an official health advisory, which would hopefully free up resources and qualify the city
for bottled water, filters, and other aid.

For the hundredth time, I wondered: *Is the official indifference
because these are Flint kids? Poor kids? Black kids? Kids who already
have every adversity in the world piled up against them?*

I was getting really mad. At times I could feel myself physically
shaking.

LUCKILY, I HAD TWO meetings on my calendar that day. At the
first one, I had a chance to see my old friend Kirk Smith, CEO of
the Greater Flint Health Coalition (GFHC), which was leading a
new community-based initiative called Children's Healthcare Access Program (CHAP) to improve healthcare access and reduce
medical costs for kids.

I was on the planning committee of CHAP, which had just gotten funded by a grant from the United Way. Our meeting was to discuss implementation. United Way runs a program called 2-1-1, through which people can obtain free information and support for essential health and human services around the clock, via phone or website. It connects people in need with critical services that save lives and support health.

But CHAP was going to take 2-1-1 a step further. Instead of just referring a child who needs a car seat to a resource, CHAP would then call that family back and make sure they got the car seat, and offer them transportation services, social work services, and enrollment in other eligible programs. But there were things to sort out. What should CHAP take on? Where should we start? We decided to pilot the program in the handful of pediatric clinics that see the most Medicaid kids in Flint. Hurley was number one on that list.

Kirk ran the meeting skillfully, raising several concerns and ideas, while the rest of us—a couple of local pediatricians, a school nurse, an asthma management specialist, and health coalition staff—chimed in. We were joined by Dr. Reynolds, who was serving as the medical director for the program, a totally volunteer position, and Jamie Gaskin, the CEO of Genesee United Way, whom I had never met before.

When the meeting ended, I got up slowly, gathered up my laptop, coffee, and phone, and made my way over to Kirk. "The water situation is not good," I said.

"Tell me more," he said.

"I'm looking at the lead levels in Flint kids' blood." Then I shared some of my research. Just enough.

Kirk and I had gone through the same public health program more than a decade ago, and he had known me long enough to know that I am never inflammatory or impulsive. But I am passionate and expressive. He was studying my face, which had become a road map of anxiety.

"I'm getting nowhere with the county and state," I said. "They keep blowing me off." While standing, I forwarded him the emails I had sent to the county and the state.

He sighed.

Dr. Reynolds wandered over and joined our chat in a corner of the boardroom. Then Jamie Gaskin from United Way came over and listened in. Somebody mentioned having seen "that water guy from Virginia Tech" on TV.

"I met with Marc Edwards," I said.

All eyes turned to me.

"He's the real deal, as far as I can tell. And the testing he's doing shows very high amounts of lead in the tap water. *Very high.*"

I could see Kirk's mental wheels spinning. Then Jamie Gaskin spoke up. He lived in Flint and had a seven-year-old son. Within a span of only minutes, he was talking about solutions: distributing bottled water and obtaining filters. Then I raised my most serious concern—the powdered formula mixed with tap water.

The only solution was getting someone to declare a state of emergency or issue a "health advisory," which we hoped would kick into gear a number of governmental responses, including bottled water deliveries, water filters, and maybe premixed baby formula. But nothing could happen without this official action. We talked about the quickest way to get it done.

I said that I was in touch with Representative Kildee's office, and a staffer there was working the USDA angle for premixed formula. What other options were there? Kirk and Dr. Reynolds were as worked up and engaged as I'd ever seen them. They were talking twice as fast as usual, interrupting each other, and tossing out new ideas at a breathtaking pace. Kirk was writing things down in his little notebook, and Jamie was channeling his frustration into action.

I could also see that I wasn't alone anymore. A team was coming together.

THAT NIGHT, AS LUCK would have it, I dropped in on a hotbed of pediatricians, the perfect place to find more supporters. The annual meeting of MIAAP was just beginning.

At the board meeting that first night, after a few opening remarks, the agenda was to discuss the dismal immunization rate in the state of Michigan and some solutions to improve it. There were about ten pediatricians on the board, including me.

I am passionately pro-vaccination, and convincing parents to have trust in the science on vaccines is one of the most important tasks facing pediatricians today. Combatting this dangerous trend of science denial, I always explain to my patients the concrete evidence about vaccine safety and efficacy—literally how many millions of lives have been saved—and I share that I have no hesitation immunizing my own daughters.

I was called on to describe a press conference that I had attended, as a representative of the MIAAP, a month earlier for a back-to-school immunization campaign throughout the state. I delivered a quick summary of that event, then remembered that at the conference I had met the new chief medical officer for the Michigan health department, Dr. Eden Wells, a sensible, no-frills woman who also directed the University of Michigan's preventive medicine fellowship. Since doctors would be talking about the importance of vaccination at the press conference, I brought my white coat, even though I am usually reluctant to wear it—or scrubs, for that matter—outside the clinic or hospital. Flaunting the white coat is something only medical students and new doctors do. But I was supposed to give media interviews that day on camera, where the white coat screams credibility. Eden told me she was sorry she didn't think to wear hers. She also noted that, in addition to the Hurley Children's logo, my coat had an MSU green S. She teased me about being a Spartan while she was a Wolverine, rival schools.

Remembering that encounter got me thinking about the health department and made me wonder again about Karen Lishinski, the agency's lead-poisoning nurse. Six days had passed since I'd reached out to her, but I still had received no reply. Lead colonized my mind again. While the rest of the MIAAP board discussed vaccination rates, I kept thinking about blood-lead screening rates. How many Flint kids who were required to get a Medicaid-mandated screening, I wondered, actually got one? The number of at-risk kids who never got screened would dramatically impact our sample as well as the findings.

The more I learned, the more I saw how wrongheaded the public health approach to lead was. It was ass backward. When we test a child for lead, we are testing the child's environment. Children become the proverbial canaries in the coal mine, as we use their bodies, their lives, as instruments to test the world around them. If they test high, that means there's lead in their environment. This is useful to know, but for the child, it's already too late.

"Primary prevention" means preventing harm from occurring before a child moves into a house, before a mom gets pregnant. A truly visionary program would be methodically identifying and eliminating the lead from our environment completely *before a child is exposed.*

As the board meeting came to a close, the MIAAP executive director asked if anyone on the board had something else to share, a concern or development. It was late. People were packing up.

I looked around. No hands were going up.

I blurted it out, my heart pounding: "*Well, we may have a lead problem in Flint.*" I was sitting in a room with the leading pediatricians in the state, after all. These were my peeps. If I had to go public, this was the place.

All heads turned to me, awaiting more information.

As briefly as possible, I explained the history of the water issue. And I knew one story would drive it home: I told them how GM had stopped using the water a year before because it was corroding engine parts, but we were still expecting the kids of Flint to drink it. I said we were looking at blood-lead levels at the Hurley Children's Clinic, and our findings were "concerning." I knew not to share numbers, or describe much else beyond saying that we were currently analyzing a larger data set.

A long silence fell over the room. I didn't want to alarm them, not yet, so I just repeated that we were still looking at the numbers, but I was concerned. I looked around the room and met the eyes of the other pediatricians. Pediatricians seem to wear compassion on their sleeve—and have trouble hiding emotions. Without needing any further explanation, they knew what was at stake. I could see that in their eyes.

Denise Sloan, the executive director, said she had just spoken with Dr. Reynolds about possible topics for our "advocacy lunch" on Saturday. He had mentioned Flint's water too—and suggested that we raise the subject on Saturday with the larger group of pediatricians in attendance.

"Sounds good," I said. "I won't have any numbers to share. But I would like to strategize."

THE NEXT DAY I attended the all-day MIAAP conference on toxic stress, and later had no memory of most of it. Sitting in a hard-backed chair in another characterless meeting room, I tried to pretend that I was watching the PowerPoint presentations on the big screen. (I call it "death by PowerPoint.") But I could have been anywhere and nowhere.

Every half-minute or minute, my phone sounded or buzzed with a new text.

I ignored the usual incoming tide of work-related stuff and pulled out anything from Kirk, Elin, Melany, Dean, Jordan, Jenny, or Marc for immediate consideration. My attention went to Jenny first and foremost.

We were furiously texting as we analyzed the data, and going over and over the design of our study. We were facing a few crucial decisions.

We couldn't decide what to do about children who had multiple blood-lead levels. For the study, we needed only one level per child. But if a child was diagnosed with an elevated level, the protocol was to take their blood levels repeatedly. Should we use their highest lead level or their first lead level? The literature seemed to contain both kinds of examples. We were constantly going back to published research, reading article after article, to see how other studies were done.

After some reluctance, we decided to ask Marc for advice—to tap his scientific expertise. He didn't seem to answer texts, but he replied to emails almost instantly. He recommended that we use the highest blood-lead level for each child. This was what he did when he looked at blood-lead levels in D.C. But more important, he emphasized that the fact that a child's first lead level may not be elevated does not "protect" him or her from future elevated lead levels.

And looking over our study design, he questioned one aspect: the time frame. We had a comparison of levels before and after the water switch. While we used the same number of months, we didn't use the exact same months of the year. And that was a problem. Water-lead levels are affected by heat and the seasons of the year. Lead in water peaks when the outside temperature is higher. That's why you're never supposed to use hot water from your tap.

What? *Crap.* Of course seasonality was a factor. While I wrestled with anxiety from the possibility that I could have gotten other things wrong, Jenny and I moved quickly. We were facing another do-over. But what would have been a painstaking, time-consuming, and months-long endeavor just five or ten years ago was a pretty quick undertaking. Due to advances in statistics software and algorithms for sorting and analyzing data, she was able to run the new study and send me the new numbers while I was sitting at a lecture at the MIAAP conference.

I thought we had it right this time—we controlled for seasons, and we used the highest lead level if a child had more than one.

I looked over the results. The number of elevated blood-lead levels was still higher after the water switch—markedly higher.

It was clear.

We had the proof we'd been looking for: kids were being harmed every day, with every sip of water they drank, with every bottle of formula.

I forwarded the preliminary results to Marc.

FROM: Mona Hanna-Attisha
TO: Marc Edwards
SENT: Friday, September 18, 2015, 11:51 A.M.
SUBJECT: Prelim results—confidential

FYI

We changed pre and post dates to remove seasonal impact and we used highest lead for duplicate leads.

EBL* % for kids less than 5

2.1% pre vs 4.0% post - p=0.024

We will soon break down to the higher risk zip codes where you saw higher water-lead levels.

* Elevated Blood Levels

Marc responded immediately:

I'm ashamed for my profession.

Marc would say those words many times in the coming weeks and months. He took the responsibilities of his engineering profession seriously and personally. Just as I knew, without any doubt, that pediatricians are entrusted with the health and welfare of children above all, Marc saw the main responsibility of water treatment experts—whether they worked at a utility, a city treatment plant, a state health department, a university, or the EPA—as providing safe drinking water, one of the central foundations of any society, from the humblest prehistoric settlement to modern nations. For him, what was happening in Flint wasn't just a Flint problem. It tarred his entire profession.

DURING A BREAK BETWEEN LECTURES, Dr. Reynolds found me—and we walked out to a corridor to take a call from the Flint state senator, Jim Ananich, who had heard news of my research.

"Talk to Dr. Mona," Dr. Reynolds said to Senator Ananich. "She's the pediatrician doing the study." Then he turned his phone on speaker so we could do a mini-conference.

"I just heard about the research you're doing," Senator Ananich said in a quiet, even tone. I didn't really know him. He was a burly, bearded guy with a reputation for being bighearted and kind. He had grown up in Flint with Flint politics—his dad was a local city official who had died at a young age.

"My staff has been prodding the state for more information about the water for months," he said. After Marc Edwards released his findings, they'd tried harder but still never got anywhere. They were told that Edwards was a quack, that he had baggage and grudges. They were told he was scaring Flint with invented numbers.

"I've met with him," I said. "And his science is solid."

I could hear the frustration rising in Senator Ananich's deep voice. "God, I am so sad to hear that. That's all the kids in Flint need, right?"

"I know."

"Can I ask a question about my new baby?"

"Of course."

"What should we be feeding him?"

"Breastfeeding is the best option," I said.

The senator responded with a long silence, then finally replied in a quieter voice, "That's not possible." He and his wife were foster parents and were now trying to adopt the baby, who was on formula. He also mentioned that the baby was stuffy with a cold.

"Get the ready-made formula," I told him. "It's premixed with water. Or use only bottled water with the powdered formula. Use a humidifier for his cold—that will help with his congestion—but be sure to put bottled water in that too. Avoid anything coming out of the tap."

As Dr. Reynolds and I walked back to the conference, both of us were fuming. He turned to ask me a question. "Have you heard about the concept of environmental injustice?"

I nodded. "Of course. I studied with Professor Bunyan Bryant."

More than twenty years ago, back when I was a tree-hugging environmentalist at the University of Michigan's School of Natural Resources and Environment, the legendary Bunyan Bryant was one of my early mentors. He was a pioneer of the environmental justice (EJ) movement—a movement that looked at environmental and public health issues through the lens of place, race, and poverty. Bryant was a Flint native—with family still there—and focused much of his research and advocacy on the city and its long history of polluting poor and brown neighborhoods. Bryant had even fought lead pollution in Flint decades ago, when a plant that burned lead-painted wood chips was built in a predominantly African-American neighborhood.

As an undergrad, I took courses, listened to lectures, and participated in workshops led by Bryant and other EJ academic groundbreakers like his colleague Paul Mohai. Bryant's work showed me how racial minorities and low-income communities faced a disproportionate share of environmental and public health burdens.

Sitting in those EJ classes, I began to see that the environmental disparities I'd first witnessed in high school weren't random. The dirty incinerator we had fought so hard wasn't in Grosse Pointe or Birmingham, affluent suburbs. It was in Madison Heights, one of the poorer communities in our county. Bryant backed up his many examples of environmental injustice with hard-core research, showing how industrial waste, incinerators, trash dumps, and chemical plants were often located in neighborhoods where residents had fewer resources to fight them.

Informed by these lessons, I dove into service learning, fieldwork, social justice organizing, and environmental health research. On one spring break, I went to maquiladoras in Mexico, where many of the auto jobs that had fled Michigan went—and saw that they were now troubled by the same pollution, poverty, and labor issues that pockmarked our history in Flint.

Bryant's work stayed with me as I went off to study medicine and public health—and, eventually, when I came to work in Flint.

He was on my mind now more than ever. In lectures, Bryant had specifically called out the persistence of lead in paint and paint dust in black and brown and poor communities as a form of "environmental racism."

Bryant wasn't one to dwell on the problems. A central tenet of EJ is that local communities must have control over their environments—and decide whether a pipeline gets a permit, or a wind turbine gets built instead of a natural gas plant. When people have a say, smarter decisions are made—both for the environment and for public health.

"Our Flint kids," I said to Dr. Reynolds, "already have higher rates of lead exposure—just like kids in Detroit, Chicago, Baltimore, and Philadelphia. And now on top of all that, they've got lead in their water to worry about."

Lead shifts down the entire bell curve of intelligence, as Dr. Reynolds and I knew, not only adding more people with severely reduced intellectual capacity but also reducing the number of exceptionally gifted people. We knew that lead is more prevalent in poor and minority communities, and thus lead exposure exacerbates our horrible trends in inequality and the too-wide racial education gap. We knew that if you were going to put something in a population to keep people down for generations to come, it would be lead.

"Environmental injustice," Dr. Reynolds said, shaking his head in disgust.

"I know," I said. "Some things don't change."

CHAPTER 15

Poisoned by Policy

OUR GOAL WAS NOW CLEAR: *GET A HEALTH ADVISORY ISSUED and alert the public.* Kirk Smith at the health coalition knew the system and he knew how to work it. And he knew that the first email should go out to Howard Croft, the director of public works in Flint. I didn't know Croft. Honestly, I didn't even know there was a public works department or what it did. But on Thursday, after our meeting, Kirk began drafting an email to him, saying there was concern that children in Flint had elevated blood-lead levels and urgent action was required.

Croft replied at 3:50 A.M.—the middle of the night, or early morning, depending on how you look at it. It was a truly remarkable demonstration of cover-your-ass and cluelessness all at the same time.

> The Director of the DEQ has issued two official statements in the last two days contending that the City currently meets all federal and state standards and we have noted that we are well ahead of our 2016 target for corrosion control. . . .

Blah, blah, blah. No urgency. No heightened concern. The rest of it was a bunch of long-winded, do-nothing bureaucrat talk. It felt like a preemptive excuse. If the bubonic plague infected the state of Michigan, should we form a committee?

The water had poison in it! The impact of lead is irreversible. They weren't using any corrosion control at the time, so Croft was lying about that—or he had bought the lies being dished out by MDEQ.

But I could ignore Croft for the moment because at nine-thirty on Friday morning, the mayor, Dayne Walling, emailed Kirk directly, wanting to talk. Croft had fortunately looped in the mayor. And now his office wanted all the information so it could "respond/act accordingly." Kirk forwarded the emails.

We were actually getting traction. Somebody was listening, even if it was a mayor who had been politically neutered by the governor. He wasn't in charge anymore, but then neither was the governor's fourth EM. That position had been recently vacated, and decision-making power had shifted to Natasha Henderson, the Flint city manager. Even so, Walling had the title of mayor, which still must mean something. He was also a famous native son of Flint who could certainly attract a lot of attention to the water issue. If we got Walling on board, things could happen quickly.

Kirk scheduled a meeting with Walling for Monday afternoon. The goal was to get a health advisory issued, which would kick into gear a number of governmental responses to the water problem—except we had absolutely no idea what authority was supposed to issue it.

That gave me the weekend to finish my analysis and create a presentation for the mayor's office. Just two days to make 100 percent sure our numbers were right. Two days. I was under the gun again.

As soon as I filled in Jenny, we hunkered down and got in touch with Marc Edwards. We were taken by his obvious care for the kids of Flint and his eagerness to help, but I knew he had another side. His reputation for brusqueness and moral certitude was legendary. He had made enemies out in the world. So far we'd seen only glimmers of this. For instance, Marc finally remembered that he had met Dean Dean before, during the water wars in Lansing, and emailed to inform us that Dean Dean had been an "unhelpful dupe" back then. But at the same time, here Marc was, being so generous and helpful to us.

Jenny and I had worked on dozens of academic studies over the years, but putting together an utterly perfect and unassailable one—in a matter of days, no less—was a bit of a leap. The pressure was intense. One minor error, even one that didn't affect the findings, would give critics the ammunition to undermine me. One minor error and all our efforts would be for nothing, and Flint kids would go on being poisoned. I was already out on a limb—and already being ignored. We had to produce a study that couldn't be.

This is where Marc was crucial. He had done a similar study in D.C. of children's blood-lead levels—and he knew what would make our study ironclad, flawless, and impervious to insult. We shared more results, how the increase was even greater in the zip codes where he had found the higher water-lead levels.

FROM: Mona Hanna-Attisha
TO: Marc Edwards
SENT: Saturday, September 19, 2015, 10:23 P.M.
SUBJECT: RE: Prelim results—confidential

FYI all is still confidential, but we also ran EBL % for kids<5yrs in the higher risk zip codes (48503-48505) pre 1.7% to post 5.3% p<0.001.
    bad.
We are really pushing for a health advisory so that we can get released necessary resources for high risk kids.

thanks, mona

He wrote back instantly. "That pretty much nails it down . . . stunning how fast you did it."

THE LAST EVENT OF the Michigan AAP meeting for me was Saturday's "advocacy lunch." This luncheon group was larger than the board meeting, about one hundred people, all pediatricians, and I knew many of them. The room was larger, another windowless conference room with round tables and so-so food.

The best thing about being a pediatrician is that caring about kids, speaking for kids, and advocating for kids is an essential part of the job description. At a conference of surgeons or dermatologists, they might have a lunch theme like "malpractice risks" or "how to maximize billing." But not pediatrics. That's why I've always felt at home in the specialty and with my colleagues. We don't just treat children's bodies—we fiercely protect their potential.

Looking over my in-box of emails with Marc, I was starting to think we were not alone. Five hundred miles away from my clinic in Flint, in Blacksburg, Virginia, there was a guy in a very different profession who seemed to care as much as we did.

As the lunch got under way, Dr. Reynolds told the crowd he'd been at home sick the week before, when I'd called to discuss the Flint water. He invited me to explain my findings to the group. With more confidence than I'd had at Thursday night's board meeting, and armed with more details, I stood and shared my concerns with the pediatricians. We were looking at the data, I said, but preliminary findings were very troubling. We were building a coalition— I'd just gotten off the phone with our state senator's office, I was working with Representative Kildee's office, and I'd involved a nonprofit health coalition and the Genesee County Medical Society. The greatest obstacles so far were the county and state health departments. We needed their data and they weren't sharing it.

As soon as the lunch was over, a crowd of concerned pediatricians descended on me with expressions of disbelief and encouragement. My chair of pediatrics at MSU, Dr. Keith English, who was sitting next to me, leaned over and pledged to help.

———

THE OTHER MARK IN my life, my older brother, had been unusually absent lately. Normally we talked every week, sometimes every
day—about our kids, our parents, what's happening in the news,
and occasionally what's going on at work. Mark has always been
wise for his age—his hair even started going gray in high school. I
counted on him for advice, certainly anything halfway legal. Whatever was going on in our lives, we knew, without talking about it,
that we had each other's backs.

A partner at the successful public interest law firm in D.C. he
helped start a few years ago, Mark is always fighting quixotic battles. Some I never heard about, some I heard about later, after he'd
won or lost. He fights on a different front, in a different place than
I do, but wherever he is, Mark is fighting the same battles that I
am—for the same people, the same ideals, the same reasons. He
battles employers to recognize low-wage workers' unions. Or he
files suits on behalf of immigrant workers for wage and hour violations. He fights discrimination. He fights racism. He takes cases
other lawyers won't, aggressively litigating in court for years and
finding ways to win. It is part of the commitment to justice our
parents hammered into us as children.

But in late August, when I first heard the news about Flint water
from Elin, Mark had been traveling, and after that he was tied up
settling a big case, representing sanitation workers who were being
cheated out of overtime pay. I wasn't sure he'd been following the
news in Flint, but knowing Mark, I suspected he was.

We had our first chance to catch up on the weekend of September 19, when he came with his family for his University of Michigan
Law School reunion. I had the medical conference to attend and
Mark had reunion events, but I knew that sometime over the course
of the weekend, amid the hubbub of a family gathering that included movies, a football game, and the usual platters of delicious
food, thanks to Bebe, we'd find time to talk. Mark's wife, Annette,

and I had been friends at the University of Michigan and classmates at the School of Natural Resources and Environment. She'd
met Mark at a party I threw. Over the years, Elliott and Mark had
become close, almost like brothers. To make things even better,
Mark and Annette's two boys are about the same age as Nina and
Layla. Our two families even spent winter breaks together, usually
with Bebe and Jidu.

A FAMILY PHOTO, AT A WEDDING, 2014

As usual, the first forty-eight hours together were almost completely about eating. Bebe had been cooking nonstop for days, making every dish Mark loved, especially spicy Iraqi curries. Elliott

planned an outdoor movie night on Friday with a projector borrowed from our neighbors, using the garage wall as a movie screen. He ordered Buddy's Pizza from the Detroit chain, my brother's favorite—half-meat, half vegetarian, jalapeños and chopped basil on top, and cooked well done. Not that we needed to order out.

On our last day of the gathering, when Bebe was exhausted from overcooking, and all four kids were busy playing, Mark and Elliott and I sat around the dining table and began working on a thousand-piece jigsaw puzzle that our friend Walter had given Elliott to pass the time during his shoulder recovery.

Mark and Elliott and I are all jigsaw puzzle fanatics, but Elliott was clearly uncomfortable. He kept moving his shoulder, trying to find a better position, readjusting the splint, then seemed to resign himself to pain.

He was the first to say something about the water. "Mark, I'm really worried about Mona. We both know she's tough, but she's been under a lot of stress lately."

"Oh?"

Mark kept his head down and was still moving the pieces scattered across the table surface and connecting some edges.

"Even the girls have noticed how distracted she's been."

"It's about the Flint water," I said.

Mark began drilling, lawyer style, for details. He'd heard about the water. He'd been focused on the high cost, and the resulting water shutoffs when residents couldn't pay their bills. Citizens went to the UN, charging that this was a human rights violation.

I sketched out the lead issues, or tried to, quickly—starting with Elin and Marc Edwards. "He's the guy who uncovered the D.C. water crisis," I said. And in another indication of how that event had been covered up—or not covered enough—Mark had only a dim memory of it, even though he had been living in the District at the time and is one of the most politically aware and astute people I've ever known. Before becoming a labor lawyer, he had been a campaign organizer and a good one.

I told him about the blood-lead-level data that Jenny and I had been studying—and ran some ideas by him, about how to make the most of it and make sure the state listened and took action.

It didn't take him long to analyze the politics. Mark's assessment: Walling was up for reelection and, as the incumbent mayor, was closely associated with the water switch. This would hinder his appetite for finding a solution. Also, Walling's limited power flowed through Lansing, so he would be hard-pressed to take a risk by going against Snyder. So it was good for Flint that Senator Ananich, who was a friend of Mark's from college, was pushing the state for answers. Mark was thrilled to hear that Ananich and his wife had taken in a foster baby—"He's going to be a fantastic dad"—and relieved to hear that Ananich had joined the group that was coming together and backing me. You'll need a good team, he said.

Then he asked about Hurley and whether the hospital was behind me. I nodded.

"It is a public hospital, right?"

"Yes. Owned by the city," I said. "Not many public hospitals are left—it's the only one of its kind left in the state."

"So if the state has gone after every single person who has tried to expose the water problem, what makes you any different?"

I had a group of puzzle pieces in front of me, all the same color. He had a point. And I just kept listening until he was finished making it.

Mark's practice in recent years had shifted. Taking a page out of Haji's entrepreneurial playbook, he had taken a risk and started his own public interest firm with four friends on Labor Day 2008, in the middle of the Great Recession. A substantial part of his work now included representing whistle-blowers in complex corporate fraud cases—something I'd never really thought about until that moment. I didn't even think of myself as a whistle-blower. So when he started talking about whistle-blowing, pushback, retaliation, and other repercussions, he was speaking from experience.

"I know you have your ducks in a row, but even if you are right, they still may try to take you down. They may go after you some-how," he said. "That often happens. Or you might lose your job and your reputation."

"I'm more concerned about the kids," I responded, "and letting everyone down."

We continued working on the puzzle. The edges were almost finished.

"She's not eating or sleeping," Elliott said, sounding emotional. I could tell he was expressing concern that had been building up inside him. Unable to talk to me about it directly for some reason, he was doing it obliquely now, through Mark. "I only wish I could do more to help."

I hadn't considered the toll it was taking on him—not being able to say anything or to help with something that was consuming me, taking up all my time. As for stress, I had to confess that it was worse than anything I'd ever experienced, far more than my wedding, childbirth, losing my grandparents, buying a house and moving, or even my dad's heart attack. It was certainly worse than medical school examinations.

"Just so you know what's ahead," Mark went on, "it could get rough. Many whistle-blowers, even if they're successful in exposing fraud, have their lives destroyed. They become obsessed, sometimes paranoid, sometimes with good reason. And it's often a years-long fight. Many are retaliated against. I have clients who have lost their homes and friends, their marriages destroyed. One even killed himself. That's why I always counsel new clients—even though they're doing the right thing—that they need to seriously consider the costs. You have to be prepared for the worst."

This made me think of the other Marc. His scars from the D.C. fight were obvious—and I wondered if he'd always have them.

And it made me think of Elin, who predicted a years-long battle.

Just then Bebe appeared in the kitchen. No doubt she noticed that we were suddenly silent. I assumed she had heard some of Mark's cautionary words, which didn't help.

"What's Mona up to now?" she asked in Arabic.

"Nothing, Mama," I said.

"Mona likes to get mixed up in things she shouldn't," she said.

"It's nothing like that, Mama," Mark said. "Don't worry. Mona's strong. *Hadeeda.*" The word means "iron" or "steel." It's Arabic shorthand for "tough as nails."

As soon as she wandered off, Mark turned to me again, knowing he didn't have much time to finish before Bebe reappeared. He spoke more softly this time.

"There are lots of risks. But it sounds like you're doing what you can to minimize them."

"Trying to."

My worst fear was that I would be humiliated in front of my parents, Elliott and the girls, my extended family, my medical residents, my professional colleagues, and my friends. That I might make a monstrous mistake and blow it somehow—and look as if I were in over my head. I worried about being the idiot who yells "Fire!" in a crowded movie theater when there is no fire. There is an Arab concept called *aeb,* or "shame." I always did my best to banish it from my brain, because it's wrong and stupid and shouldn't be in my mind, but at the same time, it had been planted there so long ago, it was like a tumor that couldn't be completely excised. It had cells that kept mutating and replicating.

Mark took a random puzzle piece from my hand and fit it into an empty spot, almost as if he were psychic and knew exactly where it belonged.

"You know what you have to do."

"Yep."

"Because this isn't about what happens to you if you do something. This is about what happens—or doesn't happen—if you don't do something."

This is what it means to be a member of a family, to have people in your life who trust you and support you and who know you sometimes better than you know yourself. Mark was there, my mom and dad were always there, and Elliott was always there, to stand with me no matter what was going down. What we had was more than love. We understood each other. We were grounded in the same core ideals and morals—and were always moving toward the same goal: to make the world more just, more equitable, and a more human place. To do the right thing, even if it was hard.

This wasn't about my career over the next few years, or even a tarred reputation. This was about how my entire life would look to me years from now, when I was Bebe's age or older. Mark knew that and was just reminding me that I knew it too. There was no way I could walk around Flint, watching kids grow into adolescents and then adults, and wonder if their problems were related to a poison that I hadn't done enough about.

It wasn't a choice, really. I was ready to do whatever was needed. And now all I had to do was try to spend the rest of the weekend being real and present and not freaked out. That didn't happen. I had to finish my analysis—and start my presentation for the mayor. But somehow we managed to finish the jigsaw puzzle on Sunday by the time Mark and his family left.

CHAPTER 16

Shortwave Radio Crackling

I F MY DAD HAD BEEN AROUND THAT WEEKEND, WE WOULD have played Konkan. We'd have been sitting around the same table, but instead of spending hours on the puzzle, we'd have been dealing out cards and eating *fistuq*, or pistachios. My dad loves an evening of Konkan. We all do. Elliott, who grew up watching his own dad play the game, can sit for long stretches at the table, hand after hand. Mama Evelyn still comes and plays—and has taught all her great-grandkids the rules of the game, the way she taught Mark and me.

And it's a truism in the family that, if we play long enough, eventually I will win everybody's money.

But my dad, Jidu, was still in China working with the auto parts manufacturer. He enjoyed his time there, as well as his second career, post-retirement, making the most of his metallurgy expertise and his engineer's fascination for problem solving. At sixty-nine, like my mom, he showed no interest in slowing down. Alloys and metals aside, he has lots of passions—for languages, international affairs, and history. He loves research most of all—digging into archives and making new discoveries.

He spent untold hours obsessing over a Persian carpet that had been passed down in my mom's family, an early 1900s Kerman rug depicting the great leaders of the world. Even when Haji was a boy, the carpet was never on the floor but was hung on a wall, in a place of honor, and featured in family portraits.

HAJI (MIDDLE) AND THE RUG (FAR LEFT), CIRCA 1920

The rug was a puzzle. Not only did my dad decipher the Farsi key on the border of the rug to identify the historic figures (all men, of course), he worked with art historians and textile scholars to unravel the rug's origin and secrets. There even was a connection between the rug and Freemasonry and the Knights Templar. It was like *The Da Vinci Code* for carpets.

Next, my dad focused on unlocking our family's ancestral puzzle—and that became his most successful side project to date. After a lot of persistence, he was able to prove that two family clans from two different villages in northern Iraq—Alqosh and Tel Keppe—had been separated three hundred years ago when one branch left its ancestral home to escape a plague. Over time they lost touch. Any awareness of a relationship vanished. But now, because of my dad's discovery—which he has given many PowerPoint presentations about—the descendants of the Shekwana and the Kas Shamoun families, literally hundreds of people who have since scattered to every corner of the globe, have connected with cousins they didn't know they had, including us.

What we never knew, until my dad's work, was our family's direct connection to a Nestorian priest named Israel Raba of the Shekwana family. He was born in 1541—my grandfather, twelve generations removed. Israel Raba and his family were famous scribes, known for their poetry and mystical literary work. They lived in an ancient monastery, Rabban Hormizd, just outside Alqosh, where they produced and guarded a library of intricately beautiful and painstakingly illustrated manuscripts.

These manuscripts were used in church services, and some were also a form of rebellion. Five hundred years ago, my family, like many Chaldeans and Assyrians—ethnically the same people—were members of the Nestorian Church, or "Church of the East," once the dominant Christian sect from the Mediterranean to India and China. As the Roman Catholic Church grew in power by converting people to its faith in the 1800s, the Nestorian liturgy and traditions were threatened. But some monasteries hid the scrolls in secret libraries—and

MANUSCRIPTS CREATED BY THE SCRIBE GIWARGIS, MY GREAT-GREAT-GREAT-GRANDFATHER'S GREAT-GREAT-GRANDFATHER, CIRCA 1700

priests transported them to safe places—so that around two hundred of the Shekwana manuscripts are still in existence today in private collections, libraries, museums, and churches around the world.

The Christian manuscripts remaining in Alqosh and elsewhere had to be rescued again recently, in 2015, when ISIS encroached on the region. In nearby Mosul, ancient churches and shrines were destroyed, and hundreds of thousands of invaluable artifacts and religious books and manuscripts were incinerated—cultural cleansing meant to weaken the soul of the Iraqi people.

MY DAD'S SUCCESS WITH this Nestorian puzzle led him to other remarkable discoveries. He stumbled onto the haunting story of Dr. Paul Shekwana, the public health pioneer who came to the United States at the turn of the last century and mysteriously died outside Iowa City. Next, he became fascinated with the life story of a relative on my mom's side, my great-uncle Nuri Rufail Koutani, who was a revolutionary in the 1930s.

Haji told my dad about his first cousin Nuri and how much he had admired him as a boy in Baghdad. Nuri was a bit older than Haji, and after attending American University in Beirut from 1928 to 1930, he studied railway engineering at MIT in Cambridge, Massachusetts. He became politically active there, in the ferment of the Great Depression, and neglected his studies, lost his scholarship, and was kicked out of school.

He returned to Iraq afterward, which he found "bleak" after his time in Cambridge—and continued his radical organizing. Nuri's *nom de guerre,* a useful device to thwart the secret police, was Anwar. He spoke four languages and was plugged into international liberation causes, including activism against the British Mandate in Palestine.

In Iraq, in 1935, he founded a leftist organization with a really kickass name, the Association Against Imperialism and Fascism. Nuri signed its manifesto, which called for independence from colonialism, and it became the starting point for the Iraqi independence movement. For an Iraqi, this is a little like an American having a relative who signed the Declaration of Independence—but Nuri's activism came with a price. He was a target of the Iraqi monarchy, which imprisoned, tortured, and killed many of his comrades. Sought by the king's secret police, he lost his job and was forced to go into deep hiding for months.

Nuri hid in a spare room of Haji's family home for many weeks, until a safer location could be found. Then Haji's mother dressed him in a woman's abaya and transported him to safety. He fled to Paris in 1937 and became involved in pre-war activism that led to the French resistance.

Then he moved on, joining 35,000 other freedom-loving idealists in the International Brigades that were fighting the fascists in Catalonia on behalf of the Spanish Republic. He served with the American volunteers in the Abraham Lincoln Brigade, was promoted to officer, and fought in the pivotal Battle of the Ebro, often called the first battle of the Second World War. Nuri was one of only two Iraqis in the International Brigades. The other was his friend, an Iraqi Jew named Setti Abraham Horresh.

What I love most about his story is Nuri's bravery, persistence, and unfailing loyalty to a borderless progressive cause. He fought for something bigger than a country or a religion, a tribe or an ethnic group. He fought for all people, for humanity, with a hope that there was another way to live. As a young woman, when I heard the stories that Haji told about Nuri, I couldn't help but imagine my great-uncle fighting with Spanish Republicans. And I thought of him hanging with Ernest Hemingway, George Orwell, Langston Hughes, and other idealists from far and wide who came together for this quintessentially romantic progressive fight of the twentieth century.

But alas, the defense of the Spanish Republic was a losing battle. With Hitler's help, fascist dictator Francisco Franco took control of Spain in 1939. It sure didn't help the cause of freedom that General Motors and other U.S. auto companies sent 12,000 trucks to help Franco, basically on credit.

Even so, Nuri continued to fight fascism and imperialism. We

NURI "ANWAR" RUFAIL,
WHEN HE FOUGHT WITH
THE INTERNATIONAL
BRIGADES

know that because—in a wild coincidence—my father's father, Dawood Hanna, the railroad station manager, had a memorable meeting with him in 1956. Along with thousands of leftists and political dissidents, Nuri was rounded up—on the order of the King of Iraq—and sent by train to a prison fortress in the remote desert of Samawah. The trains were known as *quatar al moat,* or trains of death, because most of the political prisoners did not survive. But my grandfather Dawood, who sympathized with the activists, happened to be in charge of the Samawah station. When he heard that Nuri Rufail Koutani, a well-known resistance fighter, was arriving on the train and was bound for prison, my grandfather found him and quietly offered him water and words of comfort.

It was a risky thing to do. My dad, a young boy at the time, recalls being scared that his father would be caught. Luckily he wasn't, and luckily Nuri survived imprisonment and was freed two years later, around the time of the Iraqi Revolution—and was appointed the railway minister in the Qasim regime. But not for long: a counterrevolution rose up in 1963, and imprisoning leftists was in vogue again. Thousands more were sent off to Samawah. By the time Nuri died in 1980, he had spent half of his life in hiding or in jail.

MY DAD'S CHILDHOOD WAS shadowed by the decline of his home-land. As a boy, he had been a firsthand witness to the abduction and gruesome death of Iraq's prime minister, Nuri al-Said, as he fled through the Baghdad streets on the day after the July 14, 1958, revolution to overthrow King Faisal II. A brief window of peace and prosperity for Iraq followed. And to some degree, this is the Iraq that both my parents remember fondly—a moment in their youth when their country was open, progressive, and diverse. People weren't persecuted for their religion or politics. Women dressed as they pleased. Photographs from my mom's college years in the 1960s, when she studied chemistry, look like images from UC Berkeley. She wore fashionable short skirts and held liberal values.

But unlike France, where the eighteenth-century Reign of Terror—its counterrevolution—had lasted only a few years, Iraq's terror, the Ba'athist years, lasted decades. Some of the very best people of Iraq were imprisoned and killed—writers and poets, sci-

MY MOM (SECOND FROM LEFT) ON A COLLEGE TRIP
TO BABYLONIAN RUINS, 1968

entists, freethinkers, intellectuals, academics, teachers, and political organizers. The mass murder and purging of "communists" through-out the Cold War in the Middle East uprooted the left wing, stifled secular voices, and led to the brain drain that was one of many causes of fundamentalism. The people who were dismissed and ig-nored, imprisoned and killed—or who fled—could have kept Iraq a great country, both rich and diverse. But now, after decades of tumult, war, sanctions, occupation, and civil war, their lives and work—and the ideals they fought for—are all but forgotten.

As for my dad, he had never planned to leave Iraq. But things conspired to change his mind. First, his father was arrested during the first Ba'ath revolution. Along with tens of thousands who were suspected of political opposition, Dawood was imprisoned near Basra. Worried that the police would come looking for evidence of his political views, my grandmother Mama Evelyn gathered all his books with red covers, wrapped them with a blanket, and buried them in their yard. Dawood was released from prison a couple of months later and returned home but was never completely whole again.

This pushed my dad to make something of himself. He worked hard in high school, but the slots for Christian students in Iraqi universities were limited. So he applied for a scholarship to study in Yugoslavia and was accepted.

At sixteen, he traveled alone from Baghdad to Zagreb by train, almost two thousand miles, to attend a language boot camp to learn Serbo-Croatian. He went on to the University of Zagreb to study mechanical engineering using old-school German methods, learn-ing to draft perfect letters and becoming a master of the slide rule.

On a trip home to Baghdad, he met my mom, and they stayed in Baghdad long enough to marry and have my brother. Then they moved to Sheffield, my birthplace—a steel town in England that Margaret Thatcher's union-busting policies had destroyed, much like Flint—so my dad could finish his studies. The plan was to go back home to Baghdad after that. But returning to Iraq would have meant working at a government job he'd been assigned—at

MY DAD STUDYING IN ZAGREB, 1970

the Osirak nuclear power plant—and possibly dying there. It was bombed by Israel in 1981. Instead, they moved to Houghton for my dad's postdoc work, then on to Royal Oak, when he got the job with GM.

Meanwhile the homeland my father knew and loved was quickly disappearing behind him. Almost nothing was written about Saddam Hussein in the U.S. media in the 1980s. Saddam was considered a friend of the Reagan administration, even though his brutalities and murders were well-known abroad. During the Iran-Iraq War (1980–88), the United States actively supported Saddam with intelligence and food credits—and also allowed Iraq to buy high-tech equipment for chemical weapons.

The truth is oil made Iraq a prosperous ally in the Middle East, and Saddam's anti-communist zeal blinded Europe and the United States to his true nature. This outraged my dad and fed the fire in him. He never gave up trying to right the wrongs and spread word of the horrors of the Ba'athist regime. Connecting with a community of dissidents, he increasingly spoke out against Saddam and the Ba'athists, even though my mom worried that his actions could hurt family members still in Baghdad, including her brother, who was conscripted to fight in the Iran-Iraq War. Saddam had spies and agents in the United States, people who could hurt us, but that never stopped my dad either.

My dad published opposition newsletters and newspapers, which he sent to other Iraqis abroad, and to U.S. government leaders, politicians, and other activists. He laid out the articles and photographs by hand, in his office at home. And this is where I always picture him, in my mind, when I think back on those years: holed up in his office with a cigarette in hand, listening to the crackling shortwave radio, desperately reading the Amnesty International and Human Rights Watch reports that came in the mail, hoping to keep up with the latest news of Iraq. And it was always bad news. Always depressing, troubling, even traumatizing.

I was eleven or twelve when he showed me photos from the genocide of Halabja, in southern Kurdistan. I stared at the shocking images—and he explained. The Kurdish resistance, with their brave *peshmerga* soldiers, both men and women, had been fighting Saddam for years. It wasn't enough that the downtrodden and persecuted Kurds were bombed with napalm and rockets. Saddam ramped up his attack to chemical weapons. Seven or eight Iraqi warplanes dropped bombs of poison on residential areas of Halabja.

Later, eyewitnesses gave accounts of clouds of white and yellow rising, twisting upward, and columns of smoke in the sky. The chemicals were a mix of things—different toxic gases that killed some people immediately. Others were burned and died slowly and painfully. As many as five thousand civilians were killed and another seven to ten thousand were horribly injured. The genocidal massacre is the largest chemical weapons attack directed against a civilian population in history. An entire city was poisoned.

That was the first time I saw a dead child—an infant wrapped in a pink blanket being held by her father. They had died in each other's arms. I have never forgotten it, or gotten over it. And I never will.

Mark and I grew up quickly that way. We understood that lead-

PHOTO: © AHMAD NATEGHI

THE HALABJA PHOTOGRAPH I SAW WHEN I WAS A GIRL

ers could be dangerous, that civilizations sat on the delicate edge of a precipice, and that injustice must be challenged. We were taught not to look away. That made it hard, in the ensuing years, to watch the continuing losses, the pain and suffering. Baghdad, once the pride of the Arab world—once modern, prosperous, and on the verge of freedom—quickly became a lost city in a broken country, adrift from its past.

Nightmares of Halabja persisted throughout my childhood, the trauma of the images I saw still searing and strong, but eventually this atrocity was dwarfed by our two wars in Iraq and the sanctions in between. The wars we saw on CNN. The sanctions were a quieter but more deadly affair. They hurt innocent people and never touched the heart of Saddam's operations. Overall the 1990s were a time of pointless misery, when tens or even hundreds of thousands of Iraqi children may have died. It left me wondering, throughout my teenage years and young adulthood, if the leaders of my adopted home cared about kids at all.

We knew how lucky we were. And we knew how bad things could be. Challenging injustice means standing up for the weak, the vulnerable, the abused, and the forgotten—be it in health, employment, education, or the environment. It means being vigilant on behalf of people who are treated as pariahs and scapegoats, populations that are dehumanized, displaced, and treated as disposable. It means fighting oppression at every opportunity—no matter the place or country. Mark and I have reacted to our childhoods by involving ourselves in making things better.

Elliott and I haven't raised Nina and Layla in exactly the same way. We aren't showing them pictures of murdered children but try, instead, to encourage empathy and service and a sense of belonging and identity. We tell them family stories so they know where they came from, so they know there are strong people behind them, ones who overcame struggles, even persecution, and made sacrifices to fight the good fight.

CHAPTER 17

Meeting the Mayor

MONDAY, SEPTEMBER 21. A SIMPLE PLAN: TO MEET THE MAYOR at 1 P.M., show him my study results, reveal the proof of elevated blood-lead levels, and have him issue a health advisory that would finally alert the residents of Flint to stop drinking the water.

I had high expectations for how it would go, and not just because I'm an eternal optimist. First, I was presenting numbers. It's hard to argue with numbers. Second, our mayor, Dayne Walling, was born in Flint, still lived in Flint, and was raising two kids in Flint. I felt like no matter what the political calculations were, he had a personal and intimate stake in Flint's drinking water. On top of that, I knew Walling was a smart guy with larger ambitions. He had served as mayor for six years, and while he was popular, the next election was only six weeks away. The seriousness of this sort of public health crisis on his watch—and its political implications—wouldn't be lost on him. This was a chance for him to take the water issue head-on and own it.

Yes, Walling was the same mayor who had supported the switch to the Flint River and had even gone on TV to publicize it—smiling while he drank the water. But if he had made a mistake, or been misled by the EPA and MDEQ, I hoped he would be big enough to admit it and try to get ahead of a potential scandal by coming out in favor of a real solution. As a Democrat, he wasn't in thrall to, or in the inner circle of, the Republican governor, Rick Snyder, whose administration's take-no-prisoners budget cuts were behind the water switch. From what I could see, Walling was free to reach his own conclusions and confront the state with me. The governor may have usurped his power, but he was still an important voice in the city and had the moral and leadership power of the office to represent the people of Flint and do what he was supposed to do—serve the public. Also, he was a lifelong friend of Senator Jim Ananich— they had grown up in Flint together—and Kirk Smith had only good things to say about him. I kept replaying all these positive factors in my mind—all so real and clear to me—to buoy my hope.

I wasn't just hoping for the best—I expected it. But that didn't mean I wasn't nervous. A lot was riding on me now. Kirk, Senator Ananich, and Dr. Reynolds all believed in me—and they hadn't even seen the data yet. My research and numbers had to be perfect, and so did my presentation. I wasn't a politician, government official, or trained speaker, but I wasn't afraid of the spotlight either. The days and nights I had spent in drama club back at Kimball High School were paying off.

Part of any performance is selecting the right costume. My clothing choices were limited to the things in my closet that still fit me, but I wanted to wear a suit—and found one. I had purchased it years before, back when I was a medical student and doing residency interviews. Otherwise I rarely wore it or needed it. The reason is simple: whatever outfit I'm wearing is usually covered by my white coat. It's the only thing anybody sees or remembers about me. But I rarely wore the white coat outside a medical setting. And our meeting with Mayor Walling was to be downtown, in the Mott Foundation building on Saginaw Street, where the Greater Flint Heath Coalition had its offices.

THE CONFERENCE ROOM AT the health coalition seemed cavernous, as if it were waiting for twice as many people. About a dozen of us spread ourselves around the long oval table, and I sat down at the head, so I would be next to the screen for my presentation. Dayne Walling arrived with a perfect smile. He sat down next to me, greeted me in a warm and friendly way, and said he remembered my brother from college. Like any accomplished politician, he was great at eye contact and gave off a sense of decency, confidence, and acute interest in what he was about to learn.

That same sense of openness—or even eye contact—was not happening directly across the table, where Natasha Henderson, the Flint city manager, and Howard Croft, the public works department head, were sitting. Without a word, they made known their feelings about the meeting immediately. Henderson, an *über*-professional woman, had been city manager for a couple of years already. Due to recent events, and the stepping down of the Flint EM, she arguably had more power than Mayor Walling—certainly more de jure power. She answered to nobody but the governor's office.

She was wearing her hair down, perfectly styled, and an elegant suit. She was confident and chilly when she entered the conference room with a very unenthusiastic greeting, then didn't look at me, or smile, for the next ninety minutes. Croft seemed to care even less; his sleepiness and audible yawns communicated a total lack of interest in the meeting, if not downright disapproval.

I was feeling anxious until Dr. Reynolds sat down to my right. His presence helped calm my rattled nerves. My resident Allison, who had just started my Community Pediatrics rotation, was on the other side of the mayor, sitting a little stiffly and uncomfortably between him and Senator Ananich, who was flanked by his chief of staff, Andy Leavitt. I gave Allison a warm and protective smile. I could tell she was nervous too—and would probably rather be working in the intensive care unit. Kirk was on the other side of them, which was great, because I could catch his eye and see his expression. A few physicians were scattered in other seats, including no-nonsense Pete Levine, the executive director of the Genesee County Medical Society.

Jenny was at the far end of the table, directly across from me. Good old Jenny. It felt like she and I had gone over our numbers a million times. She came as my data backup, if I needed one.

Before I'd arrived, Kirk had texted to tell me some stunning news. He'd held a board meeting for his health coalition earlier that morning, during which he announced that new data showed elevated levels of lead in Flint water. He had insisted that the community be informed. His health coalition board included Mark Valacak, the county health officer, who pushed back immediately when the Flint water was questioned. Valacak claimed it was in alignment with the Safe Drinking Water Act—and besides, no health advisory could be issued without a directive from MDEQ. When Kirk asked his board to vote on a resolution urging a health advisory, Valacak stayed silent, abstaining from casting a vote. The resolution passed without him.

THE MEETING WITH THE MAYOR got under way after some awkward introductions, which included patching in Dean Dean on the speakerphone of his car. I began my formal presentation with an explanation that under normal circumstances, this kind of data would not be offered publicly until it was published in a peer-reviewed journal. But that usually took months, and we didn't have months. The information was too urgent to sit on.

With the laptop open before me and the screen behind me, I stood while the PowerPoint slides were projected onto the screen. The first slide showed these words in large boldfaced type—in red.

PEDIATRIC LEAD EXPOSURE IN FLINT, MICHIGAN:
A FAILURE OF PRIMARY PREVENTION

My presentation was mostly science, data, and data analysis, so I started it with something human and real: a photograph of a little girl who looked just like little Reeva, one of our Hurley patients, when she was one year old. I described her as a baby who lived in

Flint. To protect my patients' privacy, she was a composite of several kids I knew. I named her Makayla, not an uncommon name in Flint.

"Makayla lives with a single mom and two older siblings on the west side of the city. Every morning her mom wakes up and makes a warm bottle of formula for her. The powdered formula comes from WIC, and it's mixed with warm tap water. Babies like warm bottles.

"She has just come into the Hurley clinic for her one-year checkup. She is smiling, looks great, and seems robust. Her physical exam and development appear to be normal. She receives her one-year-old vaccines and routine lead and hemoglobin screening. A couple of days later her lead level comes back as 6 μg/dl. It's elevated. There is no safe level, but 5 micrograms per deciliter (μg/dl) is the current reference level set by the CDC."

My next slide showed a quote from a 2012 report of the CDC Advisory Committee on Childhood Lead Poisoning Prevention: "Because no measurable level of blood lead is known to be without deleterious effects, and because once engendered, the effects appear to be irreversible in the absence of any other interventions, public health, environmental, and housing policies should encourage prevention of all exposure to lead."

I looked up at the faces around the oval table. Everybody was staring at the slides and listening intently. Even Howard Croft and Natasha Henderson were paying attention, although they had both perfected a look of politely enduring a pointless exercise. The mayor was writing in a notebook.

"What will happen to Makayla?" I asked. I didn't inflate or exaggerate the answer. Even a conservative estimate was scary enough. And just looking at the information on the screen made me angry, but I stayed pleasant—and tried to channel my passion and rage into just being as forceful as I could.

"The vast evidence supports an increased likelihood of a decrease in IQ. Even a blood level of 1 to 4 μg/dl—which is not yet considered above the reference level—drops the mean IQ by 3.7 points.

This reduces the number of high achievers, or those with an IQ over 130, and increases the number with low IQ, at 70—shifting that bell-shaped curve to the left. This impacts not only life achievement expectations but special education services and employment prospects. This has drastic implications at the population level.

"Elevated lead levels in childhood also increase the likelihood of ADHD behaviors, juvenile delinquency, and rates of arrests involving violent crimes.

"In terms of health, it impacts almost all systems of the body, the hematologic, cardiovascular, immunologic, and endocrine."

The mood in the room shifted from intense interest to concern, even despair. Senator Ananich shook his head in disbelief. He was probably thinking about the foster baby at their home in Flint that he and his wife had fallen in love with—and were now hoping to adopt. Next to the state senator, Andy, his chief of staff, was grimacing—and huffing in frustration. His cheeks were turning redder and redder. Kirk also looked shaken.

Dr. Reynolds was nodding in approval. He knew about the long-term effects of lead—he was an expert. I couldn't see Mayor Walling's face. He had stopped me twice for some clarifications, but now he was looking down at the table, perhaps morosely, maybe penitently, and still calmly taking notes.

Then I got to the costs, because this was what the city, the county, the state, and the federal government would pay attention to. People listen to dollar signs. (Wasn't that how we got into this mess?) You had to factor the number of children who had been exposed and make a forecast of costs. It was about probabilities and estimates, but that is how governments plan ahead and prepare. When there's a hurricane, the estimates of economic cost begin almost instantly. When there's a viral outbreak, the CDC and FEMA don't just make survival plans and prevention strategies, they immediately begin determining what it's going to cost to implement them.

I quoted from the latest figures I could find, based on a 2011 *Health Affairs* report by Leonardo Trasande and Yinghua Liu, "Reducing the Staggering Costs of Environmental Disease in Children," which estimated that in 2008 alone, $76.6 billion was spent on environmental diseases. "For childhood lead poisoning, it was $50.9 billion in lost economic productivity resulting from reduced cognitive potential."

Looking at costs to the state, I cited a report, "The Price of Pollution," published by the Michigan Network for Children's Environmental Health. Based on a 2009 study of a group of five-year-olds with household lead exposure—paint and paint dust, not water (because nobody had studied that yet)—it showed economic losses attributable to lead exposure ranging from $3.19 billion to $4.85 billion per year in future lifetime earnings.

Now I got to the heart of the matter: water. I cited research showing that when lead was in drinking water, the greatest impact was on pregnant mothers and developmentally vulnerable formula-fed infants:

- For about 25% of infants drinking formula made from tap water at 10 ppb, blood lead would rise above the CDC level of concern of 5 µg/dl
- Increase in fetal death and reduced birth weights

I brought it all home to the water in Flint. I referred to the work of Marc Edwards and his results from testing the Flint water. His team had provided three hundred water-testing kits to residents in specific wards or districts. Citizen participation had been incredible. A record number of those kits, 252, or 84 percent, had been mailed back to the Virginia Tech lab in Blacksburg.

Then I went into my own research. I described the methodology and approach, how we had used data processed at Hurley Medical Center, received IRB approval for our research, and zeroed in on blood-lead levels of all children five years of age and younger in seven Flint zip codes.

Two periods of study were compared: pre-switch blood-lead

levels from January 1 to September 15, 2013, and post-switch levels during the same months of 2015. The sample used was 1,746 Flint children pre- and post-switch.

Post-switch, the percentage of children with blood-lead elevation was almost double. I emphasized that this increase was contrary to every national and state trend. Elsewhere the percentage of kids with elevated lead levels had been coming down every year. Then I looked closely at water-lead levels in each zip code (based on Marc Edwards's work), to prove a causal relationship. The zip codes with the highest water-lead levels saw the greatest increases in the percentage of children with elevated blood-lead levels. To show the dramatic contrast between Flint water pre- and post-switch, Jenny and I had created a few more graphs showing comparisons.

Senator Ananich rocked forward in his seat, taking an audible deep breath. He was stunned. He leaned over to speak to Kirk and seemed distressed. His eyes were brimming with tears.

Dr. Reynolds was nodding, his mouth tight, as if fire would come out if he opened it. The mayor looked concerned and was nodding too. Henderson was still going with her stony-faced thing, but for a second Howard Croft seemed to be caving—or maybe he was falling asleep. I couldn't tell.

I went on.

The next slide compared Flint to the rest of Genesee County, where the water was the same as ever. This comparison made it clear that there was no statistically significant increase in elevated blood-lead levels in the rest of Genesee County—only in Flint.

Next, I really wanted to bring home that my blood-lead results, although striking, underestimated the potential exposure. Medicaid mandated lead screenings at a child's one- and two-year checkups, but that was way too late if you were worried about lead in water. "Infants are not screened for lead exposure," I said, and "lead has a short window of detection in blood—a half-life of twenty-eight days." Also, many children who are supposed to be screened never are.

I reiterated the concept of primary prevention, our most basic obligation—to keep children away from sources of harm. "The moment when a child's medical doctor learns of an exposure to this powerful toxin, it is too late. We grossly failed at primary prevention."

The next slide described next steps:

- Encourage breastfeeding
- No tap water for high-risk groups such as infants on formula and pregnant mothers
- Declare a health advisory
- Distribution of lead-clearing NSF-approved water filters
- Public education regarding precautions
- Reconnect to Lake Huron water source ASAP

I ended my presentation with the photograph of the little girl I called Makayla. I figured if I didn't get them with data, facts, and numbers, I could get them with emotion. Makayla looked at the camera with such trust and innocence. "She seems great now," I said. "She's smiling and looks perfectly healthy. But what will her future hold?" Leaving absolutely nothing on the table, I said, "And what does it hold for an entire generation of Flint children?"

As soon as I finished, the mayor asked a few questions. He was polite and not defensive at all. He wanted clarification about a few things and seemed appropriately concerned. "I hope this is part of an awakening," he said, "that Flint *does* have a lead problem, something I've been concerned about for years."

He continued on this note for a minute or two. Flint had even gotten an award for excellence in lead remediation, an achievement for which he felt responsible, and he was proud of it. But he was talking about lead paint and lead in the environment—he never actually said the word *water*. He referred to a public statement he had made just the week before that Flint would be "lead-free" by 2016. Huh? Now, having seen my presentation, he seemed excited that finally, *finally,* the rest of us were jumping on the lead-awareness bandwagon he'd been driving for years.

I was confused.

Andy looked noticeably disturbed. He leaned over to Jim and spoke in his ear, what I was sure was a choice profanity. They both shook their heads.

Dr. Reynolds jumped in to affirm the seriousness of the problem. That beautiful man is quite a talker, an epic rambler who goes on and on. That's his style, and it works. The room was captivated. He spoke of a crisis. He said it was an emergency. I was a young pediatrician no one had heard of, but Dr. Reynolds was a famous face and name in the city.

While he spoke, I looked discreetly around the table again, to see if this powerful soliloquy was shaking up the city officials.

It also mattered that Dr. Reynolds is African American and lives in Flint, the eighth-largest "majority minority" city in the United States—57 percent black, 37 percent white, and 6 percent other, mostly Latino. The installment of the EM, even a black one, rather than an elected official to represent Flint, had been the last straw. Black Flint residents knew how the city had lured their great-grandparents and grandparents with promises of industry jobs and prosperity, and that they had often had to accept lowly jobs, crappy housing, and segregated neighborhoods while their white counterparts were offered management positions, home mortgages, and eventually the chance to flee Flint for a nice all-white suburb just outside the city limits.

And then there was me, somewhere in the middle. I wasn't black, and I wasn't white either. I felt, as I often do, like an outsider, straddling but never completely or comfortably inside either world. My professional life now felt the same way: straddling. I work in Flint, and all my patients live here, and so do many of my pediatric residents, but I don't live in Flint myself, so I worried that I wasn't an entirely credible Flint voice. But Dr. Reynolds was. He exuded wisdom and credibility. And as pediatricians, we both had a professional obligation to speak up for kids.

After Dr. Reynolds finished, Kirk described the resolution that his board had passed that morning calling for a health advisory.

The mayor was finally ready to talk water. "We've just this morning had a meeting about this with the EPA and DEQ," he said. "And they've told us there's no corrosion issue. It has to do with old pipes in residents' homes and couldn't be helped." Something inside me wilted. He was copping out already. *Okay, maybe there is lead in the water,* he was saying, *but it isn't the city's, the state's, or the feds' responsibility—or mine. It's the people's fault.* I couldn't help but think how this echoed the lead industry's blaming of victims. Then he mentioned the prohibitive cost of switching back to Detroit water. Flint just couldn't afford that.

"Besides," he said, "isn't everybody in Flint already flushing their water?"

I pressed my lips tight together to make sure I didn't inadvertently blurt out an expletive or scream. Without visibly turning my head, I caught the eyes of Kirk, then Jenny. Jenny's mouth was agape. I could see they were as disturbed as I was.

Flushing? I was a pediatrician with an environmental health background, and even I didn't know to flush water after prolonged periods of nonuse. And did anyone think Flint residents, who were paying the highest water rates in the country, were going to waste water for five long minutes of flushing?

Howard Croft went next, attempting to back up the mayor. His job and reputation were on the line, and that was all I was hearing, or thought I was hearing. He was like Gonzo the Muppet. He opened his mouth, and no point came out.

I spoke up again. "This is a crisis. This is an emergency." I talked about needing bottled water for the infants in the city, or premixed formula. And the water had to be switched back immediately.

"That's impossible!" Henderson had been sitting at the table for almost an hour and hadn't said a word, but now she finally spoke up, with unshakable certainty. She made it clear that there was absolutely no money for Flint to switch back to Detroit water. "The pipeline to Detroit has already been sold." She shot me a disapproving why-are-you-wasting-my-time-with-this-crap look.

She reiterated the information that she and the mayor had received on their conference call with MDEQ and the EPA that morning. She believed what they had told her and was sticking to that. "They say there are no corrosion issues," she said. "The water tested fine at the source—"

"The source isn't the problem!" I interrupted, then began to describe why testing at the source was irrelevant. I had taken a crash course in water treatment only in the last couple of weeks, I explained, but I had consulted a former EPA water treatment scientist (Elin) and had met with Marc Edwards. The second I said Marc's name, I felt a blast of arctic air from Henderson and Croft.

Once everyone else at the meeting had a chance to weigh in, including Dean Dean from the speakerphone, expressing gratitude for my study and the presentation, I gave the mayor an ultimatum—delivered as kindly and benignly as possible.

It was pretty simple. If he wouldn't stand with me and make the announcement that there was lead in the blood of the kids of Flint, then we would do it without him. Kirk, Melany, Senator Ananich, Andy, and I had agreed on this step going into the meeting. Andy had been adamant about issuing a deadline for the mayor, and he wanted it to be Wednesday, September 23, at noon. I went along with it. It was an arbitrary deadline, but it gave us something to work with, and it gave the mayor's office two days to get their act together.

"We'll hold off until we hear from you," I said, "but it has to be by Wednesday at noon. We'd love to make this announcement with you."

The mayor nodded collegially, as if he were down with that. "But Wednesday might be tough," he said, suddenly reversing course. "I'm going to be in D.C. on Wednesday to meet the new pope."

Huh? I looked over at Andy and Senator Ananich, then across the table at Jenny. *Did he really just say he was meeting the pope?*

"This couldn't be more urgent," I went on, ignoring his bizarre non sequitur. "We need to get moving and declare a health advisory as soon as possible, so we can get some proper action by the feds and the state. Every day counts. That's why we need to hear by Wednesday."

Henderson grimaced, then looked away.

"This could be a win-win," I said, trying not to lose it. "We could stand together and make the announcement."

"Okay," the mayor said.

Okay, what? "We'd like to hear from you within two days, by Wednesday," I said again, in case he didn't catch it the first few times. "After that, we will share my research publicly."

"Okay," the mayor said again, nodding and beginning to rise. We all stood, said our goodbyes, and dispersed as quickly as we'd arrived. We were busy people with busy careers and lives. Some of us were focused on the kids in Flint. Some of us were looking forward to meeting the new pope.

Was he planning to bring back some holy water?

Texts began flying as soon as we left the building. Senator Ananich and Andy and Kirk believed the meeting had gone as badly as could be imagined—they were extremely disappointed. Mayor Walling was clearly of no use, we decided, and we had to assume he wouldn't be with us. We decided to plan a press conference for Thursday, September 24, without him. We still hoped he'd join us but had to start working around him. There was no reason to vilify him, we agreed, or turn him into the enemy—he had no power, being mayor in name only. The truth was that the citizens of Flint didn't have a democratically elected leader. They had a figurehead.

Over the course of the day and night, Andy called me repeatedly, sometimes conferencing everybody in, to coordinate a plan. He was a tall, relatively young guy and a political strategist to his bones—always two steps ahead, always aware of political implications, and inordinately good at swearing. A know-it-all, he loved telling us all, including his boss, Senator Ananich, what to do.

We each had a role to play. Senator Ananich was to work the legislative end. Kirk was to deal with the local media—now already circling the story after the GFHC resolution—and work with Dr. Reynolds to bring medical and health professionals on board. I was to finalize my study and presentation, while continuing to pester the state for a data set for all of Genesee County.

My tendency, when things don't go well, is to revisit my own work and see how it could be better. Was my presentation good enough, compelling enough? I also wondered about the decision to see the mayor in the first place. Perhaps I had been optimistic to assume he could be counted on to do the right thing. Maybe he and Henderson and Croft were all totally powerless—and we should have gone directly to the governor. I wondered if the meeting might have gone differently.

The rest of the day was a blur of emotion, adrenaline, texts, calls, and more meetings. When I wasn't strategizing with Andy and Kirk, I still had thousands of resident applications to sift—and the rest of my Hurley responsibilities. And I was recounting the latest events to Melany and Elin.

ELIN WAS STILL ON the case, digging into the water treatment and corrosion weeds with Marc Edwards, and what they guessed was happening at MDEQ. In the time since she and I met Marc at lunch, he and his Flint Water Study team had been continuing to work the system just as hard as we were. Marc was emailing every possible agency and government office that dealt with the Flint water or public health. He had written the mayor, the county health department, a number of people at MDEQ, and the governor's office.

Marc was already sending in Freedom of Information Act (FOIA) requests to obtain state and EPA email records, looking for answers. He had learned from his experience in D.C. to keep pushing the authorities consistently and firmly—and a FOIA request was the perfect way to exert pressure. He informed each agency that a request had been made, which alone might prompt them into proper action. But then, it could have another effect too: fear and more stalling. Rarely did the agencies turn the requests around in the time required by law.

In the clinic parking lot that evening, before going home, I called Marc to tell him about the meeting with the mayor. I was still angry and pinned the blame on MDEQ and the governor's office, calling them "Republican bastards" who put dollars above poor kids.

This was followed by a long silence.

"I'm a conservative Republican," Marc said.

"What?"

There's no way to describe the internal chaos that ensued. My mind was spinning and reeling. *Marc Edwards is a Republican?* Making assumptions and leaning on stereotypes, I had mentally sized him up as a lefty activist, perhaps with a libertarian streak. If I saw myself as committed to public health and protecting kids, Marc was probably even more so. He had devoted his career to crusading against negligence and dishonesty. He believed if you weren't on the right side of a public health crisis, you were a bystander to a crime. And he wore baby-animal ties.

I assumed this meant he was . . . well, I didn't know, but whatever it was, *he couldn't be a conservative Republican.* This news was taking a while to sink in. I knew Republicans and had grown up with Republicans. And I knew that progressives didn't have a monopoly on integrity or caring about people. At the same time, in the years I'd been working in Flint as a pediatrician, I hadn't encountered many crusaders who weren't lefties like me.

But did it matter?

On one hand, it did. Republicans had starved the government and made the decisions that resulted in the water switch—and the outcome only confirmed their suspicion that all government was bad. But on the other hand, what mattered wasn't politics or political philosophy—it was Flint's kids. That was ground I was willing to stand on with anybody. And it was a good piece of ground to occupy: anyone who wasn't on it was on the wrong side, regardless of political affiliation.

For Marc's part, I wondered if his politics made his iconoclastic activism easier—or, at least, less of an anguished struggle. I was a true believer when it came to government. I had faith in its ability to protect rights, promote equality, and mitigate historic injustice. So much of my life and advocacy rested on that. But what had happened in Flint, and what was happening with the state, was seriously eroding that faith.

TUESDAY, SEPTEMBER 22, CAME. No word from the city, nothing from the mayor's office. Every minute felt like an hour. I did my work at Hurley and tried to stay attentive and available at home, but I was a walking zombie.

I hadn't eaten a true meal, or anything beyond occasional bites and nibbles, for weeks. I was mainlining caffeine, drinking oceans of coffee. I was wired.

Once again I found myself in an antiseptic boardroom, this time to give my presentation to Pete Levine's county medical society group. A dozen doctors sat around the table, along with a few Hurley residents I had brought with me from Community Pediatrics, including Allison. As soon as I began talking, though, one of the group's members, an older physician, confronted me. Without a sense of what she sounded like, she complained that Flint residents didn't pay their water bills—and predicted this water thing would just encourage more of that, which would cause the doctor's own already high water bills to rise even more.

My mouth agape, I pretended she hadn't said that—how could she say something like that?—but Dr. Reynolds wasn't so forgiving. He got on her for her remarks, while Pete Levine masterfully kept things moving with adept political and diplomatic skills. Soon they were all rapt and on board. It didn't take anything but facts and science to convince a group of doctors. By the time I was finished, the doctors had passed a resolution pledging their support. Relieved and excited, I was starting to feel the wind at my back, encouraging me. The state and city were still stonewalling, but with my data in hand, we were building a movement of doctors.

Andy Leavitt was eager to release my study in advance of the press conference. As usual, he had a strategy, and he was sure it was the right one. He wanted to put heat on the mayor, the health departments, and the governor. I pushed back: we had told the mayor he had until Wednesday at noon, and we had to stick to that promise. But Andy doubted me, probably figuring I knew next to nothing about political maneuvering.

"I would like to scream the results from every rooftop," I texted him. Sometimes you don't have to maneuver. You just have to be honest, direct, and persistent.

Assuming the mayor would not respond, Andy then decided to reach out to *The Washington Post*. He was sure it would become a national story. (The rest of us were not convinced.) The *Post* had been late to the D.C. crisis but had made up for it with great coverage later on. If there was a major newspaper that would intuitively get the importance of a lead-in-water story, it would be them. There were *Post* reporters who already knew about corrosive water and lead leaching out of old service pipes. Andy's plan was to give the *Post* a heads-up on the Flint story—and embargo my study until immediately after the press conference. This sounded great to me. A national newspaper would be difficult for the state to discredit, and it would hit the governor's office harder.

While I waited to hear back from Andy, I got an email from Marc Edwards saying that if the mayor's office didn't come along, "it had been decided" that I'd share my results with Ron Fonger at *The Flint Journal*.

It had been decided? This caught me off guard. Why was Marc telling me this—and not Andy? And what had happened to the *Post*—and why was Marc involved and knew about a decision before I did?

My lack of food and sleep was probably a factor, or the recent discovery that Marc was a Republican, but I was suddenly paranoid and doubting his motives. *What if he has a secret agenda? What if he's just trying to get back at government because of D.C.?*

I called Andy right away. "Can you please confirm that we're going with Ron Fonger?"

"Yeah," he said. "Marc contacted a reporter he knows at the *Post*, who worked on the D.C. crisis. Word came back that the paper is busy this week with the Hillary email story. They said no. Besides, Marc really thinks Ron is great and deserves it." Fonger had been on the water story from the beginning.

"And you think that's the best approach?"

"I do."

"Okay," I said, exhaling. "Just needed to check."

Before bed, Kirk texted us all to report that Mayor Walling had called him at home, wanting to talk. The mayor had been ruminating and was full of justifications. "Everybody knows they're supposed to flush the water," he'd said. Kirk objected and pressured him to stand with us. But the mayor felt it was unnecessary.

WEDNESDAY, SEPTEMBER 23, ARRIVED, the day of the mayor's deadline. I took my mom to the hospital for eye surgery. I brought my laptop and phone and had both going in the waiting room while she was in surgery and then in recovery. Her doctor came out and, realizing that I was a doctor too, switched into doctor-speak and shared how everything had gone. The whole thing had lasted just a couple of hours. I worked on my presentation—and was in constant contact with Mark, my brother, who wanted to know first how Bebe was doing, and second, what was going on in Flint. It seemed like a very long time ago that we'd talked things out while doing the jigsaw puzzle, but it had been only a few days.

My mom was wearing eye patches and was supposed to rest. I took her photo—it was a little grisly—and texted it to my brother, proof that she was out of surgery. Then I got her into her house and into bed. I closed the curtains and dimmed the lights.

"Mona, tell me what's going on."

"What do you mean, Mama?"

"You know what I mean."

"Everything is good. I'm helping with a situation—something the state isn't doing but should."

"Like what?"

"Like it's the kind of thing you taught me to do. Something you would do, if you had a chance. My patients, my kids . . . it's like they were in an accident, a lot of them were, and it wasn't their fault."

She became quiet. I think she knew I wasn't going to say anything more than that. She had a knack for leaving me alone and giving me space when I needed it. And I knew she would ask Mark, and he would explain it better than I could.

I checked my phone. It was noon.

Nothing.

The deadline had passed.

Aeb

To make sure Mayor Walling knew the deadline had passed, somebody in Senator Ananich's press office called him to double-check. Sure enough, we were told that there would be no cooperation from the mayor's office. So we moved ahead without him.

The Hurley press people met me in my office and began working on an advisory to the media, announcing a press conference at the hospital the next day. There was almost no information in the release, or as little as possible. But behind the scenes, we were working every angle to make sure we had built a coalition of support. Andy was still quarterbacking while Kirk leaned on his contacts in the professional health and medical worlds, preparing them to stand behind me. Jordan Dickerson, our congressman's aide, was away from Representative Kildee's office but still in the loop. He planned to issue a statement as soon as the story broke. And Senator Ananich's press office was working with Ron Fonger at *The Flint Journal*, teeing him up to break the story. Fonger called, and we made a plan to meet at the Hurley clinic in the morning so I could explain my study to him—and answer any questions.

Marc Edwards gave his EPA buddies, especially Miguel Del Toral, a heads-up and told his trusted media contacts that my team had been able to produce the final piece of the puzzle that was missing in the D.C. crisis.

Blood data.

Proof of impact.

"This is a game changer," Marc kept saying.

He predicted that the state would not be able to continue stalling. He told LeeAnne Walters and other citizen scientists and Flint activists about the press conference and urged them to attend. After being ignored for so long, they had doubts that the medical community was actually going to do anything. Marc assured them this would be different. The blood data would change everything.

SLEEP WAS THE ONLY THING I needed. I had to be fresh and ready for whatever tomorrow would bring. But every cell of my body buzzed with excitement, every neuron and synapse of my brain was firing in every direction.

I was sure of my research. We had checked and double-checked and triple-checked. Jenny and I, both cautious to the point of paranoid, ran the numbers in multiple ways. They always came out the same. And although we still didn't have the state data, our sample was sufficiently large to make the point with statistical certainty—and enough proof to force the state to confront its negligence and mistakes. But there was no way to know what the reaction would be, and I was worried.

Down deep, something else was eating away at me. *Aeb.* It was difficult to describe without using the imprecise word *shame.* It was not just an Iraqi thing; it was an Arabic thing. It was the idea that you were never acting independently of your family or larger community. You always had a connection to a larger group, and there were always repercussions. If you behaved badly, or strayed even a little bit from the accepted norm, you would bring shame not only upon yourself but on your people. There was nothing worse.

Aeb is used to keep people in line, particularly kids and particularly girls. Anything could be *aeb.* Getting in trouble in school. Not going to church. Being gay. Wearing skimpy clothes. Whatever. Even small and seemingly insignificant things could be considered *aeb*. If you offered someone *chai*, for instance, and they said no, you still had to bring them *chai*. Because it didn't matter what they said—it would be *aeb* not to.

I hate *aeb*. It's a debilitating and ugly concept. I try not to conform. If somebody doesn't want *chai* or isn't hungry for dessert, I don't bring it to them. But in my mom's mind, this is *aeb*, and it drives her crazy.

Even though my family lives on the edges and outside the norms of the Chaldean community—we are progressives, political dissidents, and nonchurchgoers—the concept of *aeb* was deeply ingrained in us. My mom still talked about *aeb*. My dad did too. *Aeb* is the reason I started eating meat, so I wouldn't insult Elliott's mom with rude and ungrateful behavior. It's serious in our culture—and hard to ignore.

I thought about the press conference, the public release of my research. I thought about getting it wrong, embarrassing and shaming my family, my colleagues, my clinic, and my profession—in such a public way.

That would be the most colossal *aeb* of all.

ELLIOTT KNOWS THAT WHEN I'm under stress or am facing something big and overwhelming, I want to be left alone and not talk too much about it. So he carefully and sweetly asked if there was anything he could do. "You'll be great tomorrow," he said.

My brother emailed too, striking the balance of encouraging me while leaving me alone, something he'd calibrated over a lifetime of big-brothering. He didn't mention that he'd explained the entire Flint water saga to Bebe—how it had happened, and how wrong it was that nobody was doing anything about it. He told her that I was doing the right thing, the only thing I could do—advocate for the kids of Flint.

"But why does it have to be Mona," Bebe asked, "not somebody else?"

"Because Mona has the data. She has the proof. Nobody else has done that. She is a doctor, a scientist, maybe they will listen."

Bebe was quiet for a second. "What if somebody attacks her, or hurts her?"

That question tells you so much about my mom's life. But she didn't try to stop me, or say a word. She called my dad in China instead and told him. They studied the time difference and figured out how to tune in and watch the press conference on TV.

The last person I heard from was Andy. To him, I was just a doctor, hence clueless about how to give a press conference, handle a crowd, or be an activist. He sent a text, making sure everything was set up with Ron Fonger at *The Flint Journal.*

> ME: I talked to him. We are meeting tomorrow at Flint Farmers' Market at 8:45 A.M. for an interview

> ANDY: Excellent. Get some rest. I am sure it will be good to get this off your chest

> ME: thanks. I'll hopefully be able to sleep better ;)

I set my head down, but no part of my pillow was comfortable. No place in the bed felt right. I tossed, waiting to become drowsy.

ALMOST EVERY YEAR when Mark and I were growing up, my parents sent out Christmas cards. They did them in newsletter style, like a PR bulletin that covered all the year's accomplishments in a classic American way. In 1991, when we were sophomores at Kimball High School, Mark and I coauthored the card and wrote about two engineering awards that my dad had won and the teaching certificate that our mom was studying for. We described our own involvement in SEA, the environmental club—and the three months we'd spent organizing an eco-fair against toxic pollution with Greenpeace and local grassroots activists. We talked about our family car trip to Yellowstone, Badlands, and Mount Rushmore.

And the beginning of the card covered the latest news from the Persian Gulf War:

> Since Iraq is our homeland, we were in constant fear for our relatives. My grandparents as well as numerous uncles, aunts, and cousins reside in Baghdad; fortunately they were all unharmed. As the end of the war came, so did a little peace of mind for us, but little else for the Iraqi people still under the tormenting dictatorial eye of Saddam Hussein.

The following year the annual letter reported on family progress again. I had become the publisher of the Kimball High newspaper. Mark had gotten a summer job working for a computer company. We were both still active in environmentalism. News from Iraq was a bit better, or so we thought: "Saddam is still in power, but his end looms near, which will bring great relief to the people of Iraq."

 Happy Holidays from the Hannas

Greetings and Merry Christmas!

The Hanna family wishes you peace and happiness this holiday season. As we celebrate the close of the first year of the twenty-first century, we look forward to hearing from all of you and learning about the adventures of your lives. We, as always, have been busy at work, school and all of our activities.

The most significant change in our lives this year has been the diaspora of the family. Mark now lives in Washington, DC and Mona lives in Flint, Michigan. Mark is enjoying his post-education life. He

After 1994, my dad wrote the cards (although Mark and I edited the grammar; there are nuances in English he has never quite mastered). He seemed to really enjoy heralding our achievements. Mark and I graduated from Kimball with honors. I was president of my class, on the homecoming court. (Subversively, I wore a tie-dyed SEA T-shirt.)

But my dad focused on a commendation that Mark and I both received from the EPA, signed by President Clinton, for our environmental activism.

And his optimism about Iraq ended.

The situation under Saddam Hussein is getting progressively worse. This is very disheartening to us. After the war, the embargo has strangled the Iraqi people while Saddam is living in

AT THE HOMECOMING COURT, 1993

luxury. The suffering has triggered another exodus to the United States, especially to the Detroit area where there is an established Middle Eastern community.

By 2000, I was in medical school and living in Flint. Mark was in D.C. My mom had established herself as an in-demand ESL teacher and spent two weeks in Japan, studying the education system and culture. The World Trade towers were still standing, and President Bush wasn't yet using their tragic fall as an excuse for a new war.

Of course we continue to keep our eyes on our native Middle East. While much of the world is focused elsewhere, Saddam Hussein's brutal regime is continuing to strangle the Iraqi people and the sanctions on Iraq only exacerbate his hold on power. We pray for a just peace there and throughout the troubled Middle East.

At home, my dad was a man of Vulcan-like stoicism and restraint—he never even hugged us—but every year his Christmas tone grew darker and sadder, like the sound of a mournful trumpet. He saved the news about his homeland for the ending, a memorial space for those still in Iraq, still suffering and in danger. We never wanted to forget them. As for us, the message was clear: The Hanna family had once lived in Iraq. We loved that place in the deepest part of ourselves. And we were living with its loss—with homesickness, with disturbance, with injustice.

The loss only got worse after September 11. It hit us all, and it will go down as one of the modern world's great evil acts. And then the lies of the George W. Bush administration about Iraq's weapons of mass destruction caused more destruction, death, and suffering. Some five thousand American soldiers died in the Iraq War, tens of thousands were injured, and an entire generation of soldiers will live with lifelong trauma.

A great many more Iraqis have been harmed or killed. Since the 2003 invasion, according to the Iraq Body Count website, some 268,000 Iraqis have been killed by violence—most of them civilians—and as many as half a million have perished in total, if all the deaths resulting from the war are counted.

"Every year we mention the situation in Iraq," my dad wrote in his Christmas letter of 2004, the year Elliott and I married—and the year my father produced a PowerPoint presentation called "Mass Graves Symphony" about the atrocities in his homeland. "The occupation was bungled," he wrote. "A civil war loomed. And the final bonds between ethnic and religious groups that once lived alongside each other in peace are now broken. . . . Besides the removal of Saddam, it's hard to find anything positive to say."

In America, it's easy for some to pretend that the suffering world is a million miles away, on another planet, while we are safe and innocent. But for me, the suffering world was etched in the sadness on my parents' faces. It was written in the Christmas memos. Even though we were in the suburbs, safe and sound and on soccer teams and driving to Yellowstone, we were also connected by blood and heritage and humanity to the suffering world, and we couldn't forget it. In every picture, every radio voice, every video image, I felt the pain of Iraq. And I saw it in me, in my family, and in my children.

Now, as the press conference loomed, I was beginning to see that my family's saga of loss and dislocation had given me my fight—my passion and urgency. It was what had led me to the after-school meetings of SEA at Kimball. When I heard Roberta Magid's dismay at what was happening to the planet, I had felt it too, because I grew up with dismay and knew how wrong leaders could be, how cruel and negligent. They have to be held accountable, have to be challenged, because power corrupts, and our moral sensibility can be so dulled that we let atrocities happen right around us, unless we manage to stay constantly vigilant, sensitive, aroused, and ready to take a stand.

———

JUST TWENTY-EIGHT DAYS BEFORE, I had known very little about water treatment, anticorrosives, service lines, and lead leaching. I hadn't known much about lead exposure beyond paint and paint dust and bullets. I hadn't met Marc or Senator Ananich or Andy. I couldn't have told you the names of the directors in charge of the MDHHS or MDEQ. It was hard enough to keep track of the revolving door of EMs in Flint.

I was drawing on something deep inside me. Maybe it was the letters my mom received from Haji in Baghdad, or the pictures I'd seen of the gassing of the Kurdish babies. Maybe it was the tenacity and optimism of Mama Evelyn or the strength and integrity of my dissident parents. Maybe it was the inspiration of my heroes, fighters like Alice Hamilton, Genora Johnson Dollinger, and Frank Murphy. Or maybe there was even something in my DNA, an ancestral inheritance of persistence and rebellion and activism, handed down to me from the generations of prolific scribes who had hoped to keep Nestorian traditions alive, or from Nuri Rufail Koutani with his brave rebellion, or from Paul Shekwana with his passion for public health.

Or maybe it was simply that I was a mom. I cared about the kids in Flint as if they were my own, and I cared about injustice. That's all I knew. That's all I could think about. That fueled me, drove me on, and gave me a sense of commitment and inner purpose. All of me was in the fight now.

AROUND FOUR-THIRTY IN THE morning, I texted Jenny and asked her to redo one of the graphs. I worried the presentation wasn't grabby enough. Somewhere I read goldfish have a nine-second attention span. I didn't know the attention span of journalists, but I didn't want to just stand at the podium and show slides of words, graphs, and maps. This was not an audience of academics. I needed something more striking, more physical and real.

I called Hurley and asked a nurse on the night shift what kind of baby bottles we had. They were all small, two ounces, and contained premixed formula. That was reassuring but not what I was looking for.

"Okay," I said, "I don't want that kind. But could you put a can of the powdered formula in my office?" After I got off the phone, I realized the nurse must have thought I was crazy. Next, I sent a text to Allison. It was five in the morning, but she had a baby at home—maybe she was already up.

CHAPTER 19

The Press Conference

THE PRESS CONFERENCE WAS ANNOUNCED IN A RELEASE FROM Hurley, then Ron Fonger wrote a story in *The Flint Journal* that my study was going public at one o'clock that afternoon and my presentation would be live-streamed online.

When Governor Snyder's office heard that I was calling a press conference about blood-lead levels in Flint kids, they had an immediate, very strong reaction.

They went nuts.

They called Hurley—upset that Senator Ananich had seen a draft of the presentation already—and asked for my data immediately. I guessed that meant I was in big trouble.

Swearing and cursing was never my thing. I am a pediatrician and a mom with small kids. But now, after clocking time with Andy and Marc, who were masters of the explosive epithet, I was beginning to see the usefulness of an outrageous exclamation. Hearing that the governor's people were upset was a perfect moment to get started.

Give me a fucking break.

I had been stalking and harassing the state health offices with concerns, inquiries, and warnings for weeks. Nobody cared. *Nobody gave a shit.* Now suddenly I was in trouble for not being in touch with Governor Snyder directly? *You have to be fucking kidding.* They were trying to cover their asses in so many different ways at once, it was stunning. And somehow it was all my fault.

Another call came in to my boss, Melany, from a former lobbyist for Hurley. While on vacation out west, he had received a call from the governor's "director of strategy," who was also on vacation, asking about my study and the press conference. He was pretty angry, screaming and yelling. The former lobbyist explained that he didn't represent Hurley anymore, was two thousand miles away, and had no idea what it was about, but would give Melany a call—which he did.

Natasha Henderson was the next angry call to Melany's office. Natasha's aloof professional veneer had finally cracked. She didn't mince words. "I understand there's going to be a press conference?" She was just back from D.C., she said, and was calling from the airport. She wanted Melany to stop me.

"Well," Melany calmly explained, "it was my understanding that you were given a clear deadline, and that deadline passed. We're a children's hospital, and advocacy is our job. I'd like to think a public servant would feel the same way."

Although advocacy is a duty of a children's hospital, it almost never means holding a press conference.

Melany told me not to worry. She trusted me, and she trusted my science. If Hurley were a private hospital, or part of a hospital chain that looked only to shareholders and the bottom line for guidance, it might have quashed my sensitive research. But thankfully, it is a public hospital, one of the few left in the country, and we had a mandate to serve the community above all else. I sent my presentation to the Hurley PR people, and they sent it along to the governor's office.

That same morning another unsettling call came in, directly to me. It was Aron Sousa, the new interim dean of the medical school. Earlier in the week, when the Hurley press people had been crafting the news release, they contacted the MSU communications office to see if we could officially add MSU to the list of entities supporting my research. I had invited Dean Dean to the press conference, but he couldn't make it. I hoped somebody from the college would be there.

Aron Sousa called me directly with a response: nobody from the university would be attending.

"Okay," I said.

"The university supports you as a member of the faculty," he went on to explain, "but it cannot support your research."

What? I found the strange hair-splitting—supporting me but not my study—confusing and disturbing. I worried I was being thrown under the bus already.

All along, Dean Dean, who ran the public health program at MSU, had been supportive but had also warned me that this was the health department's job, the state's job, not mine. I had argued with him that advocacy is part of my job—and the people at the health department were clearly not doing theirs. I told Aron Sousa the same thing. It's my job—and also the medical school's job. Our mission as a community-based land grant medical school was "Service to People." I don't think they were actively trying to discourage me. It was more nuanced than that. Dean Dean wanted me to understand that I was stepping out of my lane and into a role that would threaten people. He was a military man and probably saw things in terms of a rigid chain of command. I was stepping out—but what do you do when the people at the top of the chain are doing nothing and every day counts?

I think Aron was just caught off guard—and didn't have much information about what was going on. His initial response, before he digested the whole story, was the classic risk-averse turtle move. Duck into your shell. And maybe it would have been easier for him and everybody else if I did that and stayed a good little pediatric residency director and waited for the proper authorities to come to their senses.

But that's not how I see things.

The world shouldn't be comprised of people in boxes, minding their own business. It should be full of people raising their voices, using their power and presence, standing up for what's right. Minding one another's business.

That's the world I live in.

And that's the world I want to live in.

I wondered if part of the pushback was about the $270 million the state government approves for MSU every year. The power of money can't be underestimated, ever. As Karl Marx said, "It transforms fidelity into infidelity, love into hate, hate into love, virtue into vice, vice into virtue, servant into master, master into servant, idiocy into intelligence, and intelligence into idiocy." Or as the Bible says more succinctly, the love of money is the root of all evil.

There was definitely no way to remove the MSU logo—a Spartan *S*—from my white coat, where it was embroidered in bright green on the other side of DR. MONA HANNA-ATTISHA. Even if the university didn't support the words coming out of my mouth, this doctor would be wearing her white coat.

I CAME INTO THE clinic early and prepared for my meeting with Ron Fonger, the journalist from *The Flint Journal* and its online version, *MLive*. His careworn expression made me suspect he was as stressed and disturbed as I was. The guy had been beating the drum for a year or more about the Flint water—and had written hundreds of articles by that point. I went over my study and data with him, with his agreement that the information would be embargoed until the press conference, but when it came time to publish an article, he would be ready.

Before I left the clinic, Allison found me. In response to my text earlier in the morning, she handed me two of her son Liam's baby bottles. They were both clear. One was bigger and easy to see. I took it, thanked her, and put it in my bag.

Afterward I returned to my Hurley office to polish my presentation, looked at the slides for the millionth time, and fiddled with the new graphs that Jenny had sent over. I found another pediatrician to cover me at the clinic that afternoon. Then I sent an email to all my residents and faculty inviting them to the conference. No matter what, this was going to be educational.

I took Liam's baby bottle out of my bag. I walked out into the hallway, to a small bathroom across from my office. I filled the bottle with Flint water. It looked okay, pretty clear. But that wasn't the point. The point was what our eyes couldn't see.

THE HURLEY CONFERENCE ROOM, just a few floors down, was ready by the time I arrived to load my presentation. Just a few days before, I had given a talk to the faculty there and faced a room of familiar faces. Now there were forty or fifty people there, mostly reporters, photographers, and TV camera crews, with their spots staked out and their microphones stuck to the podium. More people were coming through the door behind me. As I walked closer to the podium, I recognized the dogged ACLU reporter Curt Guyette from his online video interviews, and thanked him for his solid reporting.

Kirk Smith was standing near the door. We shared a look of astonishment. He walked over to tell me that Mark Valacak, the head of the county health department, was planning to attend. At my suggestion, Kirk had called him the night before and urged him to come. From my earliest emails with Valacak, and his noncommittal behavior at Kirk's health coalition meeting, I didn't think he understood the gravity of the situation. But I thought the guy deserved a chance for redemption. And having the county health department represented at the press conference would be good for us—and for them.

Across the way, I saw Natasha Henderson in another perfect suit. I was surprised when she made eye contact. That was a first. I was even more surprised to see Howard Croft.

Were they planning to refute my work—and argue with me?

Deeper into the gathering crowd, I caught sight of Karen Weaver, who was running against Dayne Walling for mayor, as well as city council members and various Hurley board members. I recognized LeeAnne Walters, the tough Flint mom and military wife who'd started it all—who'd gone to town meetings, called the EPA, tracked down Del Toral, and gotten in touch with Marc Edwards. Her life over much of the last year had been about the water and was given over to activism on behalf of her kids and all Flint kids. I introduced myself, thanked her, gave her a hug, and could tell after just a minute of talking that she was a natural leader and fighter. Someone you didn't want to mess with. She'd had struggles in her life and had come out stronger.

Marc was in Blacksburg and planned to watch online, as did Elin, Jordan, Senator Ananich, Andy, Elliott, and my parents. Allison was at the clinic, hoping to see the press conference while she was on duty seeing patients. They had all contacted me earlier in the day and sent texts and emails of support.

People were still arriving, now about one hundred in all. They were mostly media and hospital employees and maybe a dozen or more activists—the tireless "water warriors." As I looked out into the sea of cameras and microphones, my residents and medical students were easy to spot in their white coats.

AT THE PODIUM WITH (FROM LEFT) MARK VALACAK, CLARENCE PIERCE, PETE LEVINE, KIRK SMITH, AND DR. REYNOLDS, SEPTEMBER 24, 2015

The podium was way too tall for me, and I could barely see over the microphones stuck to it. As a demonstration of support, we decided that our team would gather around me while I presented. Kirk, Dr. Reynolds, Pete Levine, and Jamie Gaskin from the United Way stood behind and next to me; so did Clarence Pierce, the CEO of Hamilton Health Network, a nonprofit clinic for the underserved in Flint, and even Mark Valacak. All men, all much taller than I am. I felt like I was surrounded by bodyguards.

Just before starting, I looked directly across the room and settled my eyes on Crystal Cederna-Meko, my associate residency program director, a pediatric psychologist, and a good friend. She would be my anchor—a friendly, assuring face to focus on.

I talked for about forty minutes, which went by in an instant and felt like a lifetime. It was pretty much the same presentation I'd given the mayor, but with some clearer graphics and a punchier, more practiced delivery. I knew instinctively I was doing okay because the room was totally silent, hanging on every word, every number, every blood-lead level.

Amid the science and facts and research numbers, I kept to the story of Makayla. The human aspect was critical. But it wasn't enough. I needed to open their eyes—*to make them really see.* I put the clicker down, and, carrying Liam's baby bottle filled with Flint water, I walked to the head table and tried to open a can of powdered formula. My plan was to mix the powder with the tap water while the room watched. But the formula can was difficult to open and then, once opened, very messy. So I improvised and just held up the baby bottle filled with Flint water.

"This is what our babies are drinking, for their first year of life. Lead-tainted water during the period of most critical brain development."

Usually I talk fast, but I needed to hold that moment—to slow down the delivery. I paused and held up the bottle for a few seconds longer.

I wanted everybody to really see it.

It needed to sink in.

I PAUSED AND HELD UP THE BABY BOTTLE FOR A FEW SECONDS LONGER

I went on, from one slide to the next, all the way to the list of recommendations for Flint residents—no tap water, lead-clearing filters, breastfeed, breastfeed, breastfeed . . . and the "impossible" one, switching back to Great Lakes water. I could hear my own voice as I spoke: calm, loud, clear, and assured. But behind the podium, under my white coat, my heart was booming in my chest like thunder. I wondered if the microphone could pick up the sound.

Dr. Reynolds spoke after me, reiterating lead's harm and issuing a call to action. Then the media began asking questions. Some were for me, like "Who funded your research?"

"No one."

"How big was your research team?"

"Just me and another young mom."

A puddle of reporters encircled Mark Valacak, the county health officer, digging into fine points and specifics. *How many children were impacted? How long did the city and county know?*

The water warriors were visibly angry and sad, exhausted from fighting for well over a year already. And from being attacked, dismissed, and ignored. It made me mad to think about. LeeAnne Walters was wiping away tears. Curt Guyette raised his voice in exasperation and called out from the scrum around Valacak, "You are lying, sir! You are lying!"

I couldn't hear what Valacak had said, but he looked dazed and shook his head. Another group of reporters surrounded Natasha Henderson. I couldn't hear what she was saying, but she seemed poised and ready with a response.

I looked around for Mayor Walling. He wasn't there.

FOR A FEW MINUTES after I stepped from the podium, I was buzzing with a postadrenaline high. The stress of anticipation was behind me. The press conference was over. The news was finally public, released from my cycling mind and heavy shoulders and out in the world. For the last month, every second that had gone by without an announcement about the water was agony, knowing Flint kids were still drinking it, Flint babies were still being fed formula mixed with it. It was an awful secret to hold inside me—and it had taken a toll.

I had done my job. My science was right. And people kept telling me how good I was. *You're awesome! That was amazing!*

Melany texted. "Excellent job!"

Congratulations came in from Andy and Senator Ananich, who were freaking out about the baby bottle filled with Flint water. "What a visual!" Andy said. "Nobody will forget that. I know I won't!" He sounded totally amazed, as if he had underestimated me, thought I was just a doctor. But deep inside it seemed that I had been preparing for that moment my whole life.

Those first few minutes of euphoria were wonderful; I'd later wish I'd held on to that feeling for a little longer. But it didn't last. The truth was out, but—just as I'd been warned—it would awaken new enemies.

CHAPTER 20

Splice and Dice

THE BLOWBACK BEGAN IMMEDIATELY. THE CITY WAS WAFFLING toward some kind of concession, announcing it would have its own press conference the next day. But the state of Michigan dug in harder. Even before bothering to analyze my findings, the governor's office and the state agencies launched a systematic effort to undermine and discredit me.

Brad Wurfel was first, doing his job as the spokesman for the corrupt and mysterious powers at MDEQ. He repeated his familiar refrain: Flint water was within acceptable levels. *Everything is in compliance. Everything tests fine.*

There was simply nothing to worry about.

Then he focused on me. The guy seemed to have one speed, one method—attack and destroy. Just a couple of months before, he had gone after Miguel Del Toral, calling him a "rogue employee" of the EPA. Then he had gone after Marc Edwards, saying he was "fanning political flames irresponsibly."

Now it was my turn. As soon as the news conference ended, his ugly statements began popping up online—not just in the local outlets but, by the next day, in state and national media. My conclusions weren't just "irresponsible," Wurfel said, dripping with condescension. "I would call them *unfortunate*.... Flint's drinking water is safe in that it's meeting state and federal standards."

I was an unfortunate researcher.

Meanwhile the state health agency, the MDHHS, fired off a scientific-sounding statement that my findings had been due to a "seasonal anomaly"—a refutation they hoped no reporter would have the temerity or science background to dig into. The problem, they said, was "seasonality." But I had factored for that—Jenny and I had controlled the study for seasons. The MDHHS didn't even ask or appear to want to know that. It was an obvious attempt to confuse, distort, and dismiss.

I DON'T REMEMBER ANYTHING about the drive home. My head was spinning, my phone was buzzing. I took repeated calls on speakerphone.

I skipped the early dinner that Elliott had made and instead got ready to attend Layla's back-to-school night. Her school calls it Curriculum Night, and I never missed one. It was important to be there, not just for me and for Layla's teachers but for Layla. She was becoming increasingly vocal about my absences from her life. There was no getting around the fact my focus on the water crisis was taking an emotional toll on the family. Elliott had the benefit of understanding what was going on, but even he said the house felt empty. Meals came and went, but I was never at the table. Even when I was home, my eyes were on my iPad or laptop or phone— and my mind was elsewhere.

Layla being Layla, she wasn't going to let this slide. Her voice was the loudest and most persistent, but she was speaking for Nina and Elliott too. At night, she often came into our bedroom, wanting to sleep with us, pressing her little body close to mine. And when she saw me looking at my laptop, she said pointed things like "I know your work is important, but I'm important too. Can you please turn off your computer?"

I wanted to be a good mom, but as hard as I tried, it was impossible to focus on anything but Flint water. The stream of emails and texts was incessant, and I couldn't shut off my rage and frustration at the officials in charge.

Just before I arrived at Layla's school, Andy and Senator Ananich called—they were driving back from Lansing to Flint together. They told me they'd just spoken with Nick Lyon, director of the MDHHS, on speakerphone. Lyon had pushed back about the water and lead, my study and the numbers, and mentioned "seasonality." Senator Ananich urged him not to reflexively discredit the data or "the doctor." The sensible approach would be for the health department, which had a staff of epidemiologists, to check the numbers themselves. Lyon assured him—yes, yes, of course, there was no point in going after the doctor.

But just as Lyon was reassuring Senator Ananich over the speakerphone, they heard the voice of Angela Minicucci, the MDHHS press person, on the car radio. She was giving a public statement. My results were "not consistent" with the state's data, she said. A spokesperson in Governor Snyder's office, none other than Sara Wurfel, was claiming that the Hurley data had been "spliced and diced."

Spliced and diced? There's nothing worse to say about a scientific study—or about a scientist. Splicing and dicing meant that I was knowingly lying. Back in my eighth-grade science class, Ms. Eisenhardt had drilled us over and over about the "scientific method" and how scientists are supposed to test their hypotheses. Manipulating my data to get the result I wanted would have been the biggest scientific sin. This accusation wasn't just a punch in the gut. It felt like a public stoning.

This implicated another scientific sin, one I was nakedly guilty of: my research had not been peer-reviewed, which was highly unusual. Peer review is both an ancient and modern procedure, a way to legitimize research by independent experts prior to publication. But peer review can take months. So I knowingly skipped that step.

It was academic disobedience and a risk to my reputation. But urgency called for it. The research was too important to wait another day.

I arrived at Layla's school. As I was sitting in her second-grade classroom, in one of those tiny chairs, I looked at the screen of my phone under the desk. Senator Ananich and Andy were working every possible angle, defending me against every state agency that was attacking me. Representative Kildee was arguing with the EPA about my findings. Kirk was texting me, horrified by the vehemence of the state's counterattack. Jenny, in her very data-driven way, began constructing statistical counterarguments.

Under the school desk, my arms were trembling, and my hands were shaking. My heart was beating in my chest at such a rapid pace that my FitBit bracelet recorded two hundred beats per minute. (I never wore it again after that night.) I told myself I was still wound up from the excitement of the press conference. I needed to focus on Layla's teacher, and listen, and be there 100 percent, like all the other moms and dads in the room.

But the media had found me; the onslaught was beginning. Outlets wanted copies of my study and my reaction to the MDEQ and MDHHS backlash: "What is your response to the state? They are saying that you are wrong."

What could I say? My study was meant to speak for itself, to stand on its own. Now I began to doubt it would. The prospect took me to a very dark place. *What if I told the truth and no one listened?*

Feeling defeated, I left Curriculum Night early—and left Elliott to meet Layla's teachers without me. I drove home, went upstairs, and curled up into a ball on the bed. I was scared and sick and not sure what I would do next, except probably throw up.

More was coming. I knew that now. The state would not revisit its own science—I couldn't expect a decent, responsible reaction or governmental accountability. Instead I was going to be forced to defend my work and expose the problems with the state's denials. The onus would be on me to keep fighting.

You don't necessarily hear this part of the story often, but when you're in the middle of a backlash, the psychological stress is extraordinary. The emotions are big, overpowering. And they come as a total surprise.

Yet in Flint, everybody else who had gone up against the state had been attacked, so why was I so stunned? My brother, Mark, the whistle-blower lawyer, had warned me this could happen.

A rush of doubts began to overwhelm me.

The state employs dozens of epidemiologists with decades of experience. They were looking at their data, exactly the same kind of data as mine but more of it, and they were telling the world I was wrong. What if I was?

I worried that I'd thrown everything away on a hunch.

I worried that the results weren't peer-reviewed.

I worried that I was in way over my head.

I worried, most of all, that I was doing exactly what the state accused me of: creating a problem where none existed and adding anxiety to the already stressed lives of my patients and our city.

Elliott walked home alone from Layla's school that night. I heard him enter the house, then rush upstairs. As soon as he found me, he tried to comfort me and even to coax me out of bed.

I broke down, sobbing. "I just told all of Flint to stop drinking the water and that all their children are being poisoned."

"You did the right thing," he said.

"I don't know."

"The kids of Flint needed you at that podium today."

"I don't know."

He tried to distract me. "*Scandal* is premiering tonight," he said.

I had been waiting months for the new season—I'd even put the date down on the calendar.

"Come on," he said. "It's just about to start. It might help get your mind off things."

"I don't care."

"Mona."

I said nothing.

"*Mona.*"

Nothing.

I felt defeated and so unbelievably small. In the coverage of the press conference, the media referred to me as a "local pediatrician." Not a scientist, not an authority, not an academic. *Local pediatrician.* It was true. That's what I was. But every time I read those words, I felt smaller and smaller, like a clueless quack, a know-nothing. What had I been doing holding a news conference? I had wandered out of my league—and was now being ridiculed. I couldn't imagine feeling good again, ever.

—

CHAPTER 21

Numbers War

SOMEHOW I FELL ASLEEP, A COLLAPSE FROM TOTAL EXHAUSTION. A good six hours passed, and I barely stirred. At three in the morning, I woke up. The house was dark and quiet. I sat up and took a deep breath. I found my glasses and put my hair up in a knot. I grabbed my phone and laptop and went downstairs.

A new feeling was growing inside me. My despair was being replaced with something else—a sense of strength and certainty. This fight wasn't about me. It had nothing to do with me. It was about Reeva and her sweet smile, that look of trust on her face. It was about the way Nakala's fuzzy head felt in my hand.

And it was about Jalen, who had pinkeye. And Macy, who told me at her four-year checkup that she wanted to be five years old.

And Tyler, who sprained his ankle.

And Sasha, who always asks if she can use the stethoscope to listen to my heart.

And Brandon, Jasmine, Chanel, and Nevaeh.

All the Flint kids I knew and saw, all the kids that I'd ever known and seen, were pushing me forward, lifting me up. This war of numbers and data was really about children. Each number on my spreadsheet was a child, a patient I'd seen and cared for. I knew their faces. I'd patted their heads, held their hands, hugged them tight. And for them, I'd keep fighting and never stop. Sure, I was a local pediatrician. But I was also a scientist, an advocate, and now an activist. Nothing was going to make me back down. Let the state "experts" come after me, try to destroy me with *all their lies.* They were scared and negligent and arrogant. They were incompetent, not me. In less than a few minutes awake, I was in high gear, making plans for my next move.

OTHER THINGS KEPT ME GOING. Earlier in the night, Bebe had called. I prepared myself for the kind of backlash only my mom can deliver. So I didn't let on how I was feeling or say anything about news stories or things being said by the state. I wondered what she knew but pretended everything was just fine and I wasn't upset or troubled.

"*Shlonke?*" It means "How are you?" in my mom's soothing Baghdadi colloquial. Non-Iraqi Arabic speakers don't understand it and sometimes make fun of it, but I find it comfort food for my ears.

"*Zayna.*" "I'm fine." Oops. Not my usual "super" or "awesome," but surprisingly she didn't catch on.

"I watched the press conference on TV," she said, then went on to say that my father woke up in the middle of the night in China to watch online. Afterward they Skyped and talked about it.

I waited for that sound to come into her voice, the fear and panic, the *aeb.*

I waited, but that night it didn't come.

"Mona," she said. "You were so brave, *mumtaz*—excellent."

IN THE DARKNESS OF early morning, the room glowed from the laptop screen. I began to fight back.

Science was on my side. Numbers don't lie. Jenny and I had checked, double-checked, and triple-checked our findings. We had scoured the literature and replicated previous studies. We had controlled for seasons, used the appropriate age range, removed duplicates, analyzed for the highest lead level for each kid, and just to be sure, anticipating pushback, we had analyzed the first lead level too.

No matter which way you ran them, the numbers came out the same. The conclusion was the same: they showed an increase in the percentage of children with elevated lead levels. But we knew that even this conclusion was a gross underestimation of exposure. The lead levels across the nation, the state of Michigan, and the city of Flint had been decreasing for the last several decades. An increase was highly unusual—an anomaly. If the percentage of children with elevated levels had stayed the same, or plateaued, it would still have been a red flag. But this was much worse.

As part of the state's counterattack, it released a tally sheet of total numbers of children in "City of Flint" zip codes with elevated lead levels by month for the past few years. It was a one-page PDF—not the raw numbers I had been pleading for—and was meant to be "proof" that I was wrong. It was basic and crude. The tally sheet showed monthly counts, used a large age range (up to sixteen), and looked only at first-time lead levels. At first glance, it didn't show an increase in the raw numbers of children with elevated blood-lead levels. But honestly, it was so unsophisticated and unscientific, Jenny and I didn't know what to make of it. There were no graphs or statistical analyses, which you would expect for something like this.

Pulling my thoughts together, I quickly wrote an email to my team and to Ron Fonger, articulating why my science was solid and why the state's numbers were different. Our studies were not even comparable, in an apples-to-apples way. And even if the state had done its analysis the same way and found numbers that were still not consistent with mine, it didn't mean my numbers were wrong. Expecting that we might be in for a years-long fight, as in D.C., I tried to look ahead. I wasn't going to give up on my faith in good science, but my focus had to be on marshaling irrefutable evidence of harm.

We were in a numbers war.

But the numbers were kids—real kids.

Suddenly an idea came to me—a way to prove the state wrong and at the same time improve my own work. It involved the use of zip codes. For our study, Jenny and I had discovered that even if a child had "Flint" in her address, it didn't necessarily mean that her house was receiving Flint water. Flint zip codes included kids who were not getting Flint water, and if they were included in a study, it diluted the harm.

What if we looked at the blood data again, I thought, using geographic information systems (GIS) software to sharpen the exact neighborhoods in the city where the water was received? We could create maps of Flint—like John Snow's maps of his London neighborhood during the cholera epidemic! And if the blood-lead levels of children in neighborhoods that only received Flint water still tested higher, the impact would be even more conclusive.

As an undergraduate in environmental science classes, I had worked with GIS. But that was long ago and the software had evolved and improved. I wondered if my friend Rick Sadler, the nutrition geographer at MSU's public health program in Flint, could help us. He usually maps grocery stores in urban areas to help identify "food deserts," places where there was nowhere to buy real food. I knew he'd have the latest GIS software down cold. I emailed him, and he quickly agreed to help.

THE WORDS "SPLICE AND DICE" were repeated throughout the next day and night, as the news cycle continued to run Flint water stories. Later on, in studies of the media coverage and its impact on the Flint crisis, Thursday, September 24, would mark a high point. The state showed no signs of backing down or even checking its facts. But good things were starting to happen too—important shifts, small and large.

On Friday morning, the day after our press conference, the mayor's office issued a "lead advisory." I had no idea what that meant, but it was a good start. Finally, people would be informed. And I was thrilled to hear that the brand-new superintendent of the Flint Community Schools had unilaterally decided to stop allowing the kids to drink the water in school. When it comes to protecting kids, pediatricians and educators are almost always on the same page.

In constant communication with Jenny and me, Rick began geomapping the kids with elevated blood-lead levels in Flint—and within a day, we had new findings. The more precise mapping technique increased the percentage of kids with elevated lead levels to more than double. And it allowed us to pinpoint hot spots in certain wards and neighborhoods with the most elevated blood-lead levels. Any doubts vanished, but while it was gratifying to know my work held up, it was also devastating.

The same day, I spoke with Kristi Tanner, a statistician with the *Detroit Free Press* who had been assigned to review the tally sheet released by the state. I shared my study with her, describing how we had carefully designed the work, factored for seasonality, and still found rising blood-lead levels.

Tanner and reporter Nancy Kaffer reviewed my research and did their own analysis of the state's ridiculous data release. The next day the *Free Press* ran their story. It not only backed me up, it called for action. This time, rather than referring to me as a "local pediatrician," I was a "hospital researcher."

> Data that the State of Michigan released last week to refute a hospital researcher's claim that an increasing number of Flint children have been lead-poisoned since the city switched its water supply actually supports the hospital's findings, a *Free Press* analysis has shown.
>
> Worse, prior to the water supply change, the number of lead-poisoned kids in Flint, and across the state, had been dropping; the reversal of that trend should prompt state public health officials to examine a brewing public health crisis.

Yes, you read that right. The state's single-minded mission to distract and deny had caused it to miss the fact that its own summary data—made public—confirmed my study rather than un-dermined it. The newspaper ran a hard-hitting editorial by Kaffer that took on Governor Snyder directly: "It's hard to understand the resounding yawn that seems to have emanated from the governor's office, following news that an increasing percentage of Flint kids have been lead poisoned after a switch in the city's water supply."

The *Detroit Free Press* carried weight, and people would pay attention. Would Governor Snyder?

WHILE WE WAITED AND pushed and insisted, our first priority was to get clean water to the kids and their families. I shared my updated analysis highlighting the hardest-hit neighborhoods with other members of my team, and they made those places our priorities. They began grappling with logistics of water and filter distribution while I was consumed with my Flint kids.

What would their future hold?

I tried to see down the road and consider what they would need. Thinking of the Flint sit-down strikers and other effective organizers, I decided we needed to compile a list of demands right away. I had learned that tactic back in Roberta Magid's SEA club. A list of demands gives your advocacy mission focus and direction. Otherwise, you could waste all the time and energy you spent getting to the bargaining table.

Where did we want to wind up? What should we hold out for?

A list began appearing in my mind. There is no magic pill for lead exposure—no antidote, no remedy, no time-honored or easy solution. The treatment is prevention. With lead, public health people strive for primary prevention, but we were way beyond that. So the next step was secondary prevention, which is what you do when a population is already exposed to something but you want to limit the impact. Lead may have already harmed Flint children, but there were things we could do to help them.

Fortunately, in the last decade, there's been a lot of research on how to deal with the lasting impact of toxic stress. Some of the advances came from technological breakthroughs in neurology and the study of brain development. Other advances came from work in the fields of trauma and economics.

To put it as simply as I can: early trauma and toxic stresses leave a mark on the brain and change neural pathways. Children exposed to adversity need to be soothed, loved, and taught how to cope and be resilient; they need to be properly nourished and surrounded by people who value them; they need policies that support them. With all these in place, they can cope and rebound; otherwise they may be living with the impact forever.

And then it hit me. Lead was only one of many developmental obstacles that our children faced. We had to frame their population-wide lead exposure, the entire trauma of the water crisis, as an additional toxic stress to a community already rattled with toxic stresses. This would give us a playbook of sorts. Our knowledge of the plasticity of the brain, its ability to rebound despite adversity, gave us hope. There were things we could do, interventions that were known to mitigate toxic stress. We had a scale we could tip. The burgeoning neuroscience in this area gave us the answer: the treatment was to build resilience.

There were tested ways to improve a child's chances of a healthier life by investing in them, especially the youngest and most vulnerable—early education, family support, proper nutrition, school health, and quality childcare. We needed what experts were now calling an "ecobiodevelopmental" approach, with short-term, intermediate, and long-term interventions that touched every level of a child's life. Science had been ignored and denied in Flint, but it was science that was critical to uncovering the crisis. Our science spoke truth to power. And now I wanted the science of child development to lead the way in our recovery.

Kirk and I began formalizing a list of demands for Flint kids. I wanted a long list. I wanted a complete one. I wanted the most extensive and proven-to-work interventions available, regardless of expense. My Flint kids deserved this and more.

———

I KNEW INSTINCTIVELY THAT I had to be strategic about the media. Almost as soon as I stepped away from the microphone at the press conference, I was drowning in interview requests. Sure, I could say the exact same thing every time. But what would have the most impact? Doctor-activist Alice Hamilton once again became my role model. I was going to make my case as loudly and clearly and scarily as possible.

Doctors are trained to be calm and comforting. But this wasn't a time for that. Now that I had a microphone in front of me, I was going to talk about the truly scary things we were facing, with no sugarcoating. Neurotoxicity. Epigenetic or multigenerational effects. Permanent cognitive impairment. Criminal behavior. All these were entirely preventable, literally man-made.

I was going to talk about the GM engine parts that corroded. I was going to talk about the Flint residents and activists who had been dismissed and ignored. Over and over again, I was going to say, "There is no cure." It was all true—and had the side benefit of being shocking and terrifying. I wanted people to hear me on the radio and see me on TV and come away as angry and freaked out as I was.

Every day I had at least one interview. One day I knocked out fifteen—on TV, radio, and print. It was exhausting. Some days all I did was talk, talk, talk to reporters. It got to the point that Elliott and the girls saw more of me on TV than at home. The Hurley press office was overwhelmed with requests, and they tried to do their best but couldn't keep up. They were used to preparing for one media event a month, like a prostate-screening fair. So I began organizing much of the media end of things myself.

There were stories, and stories about the stories, most of them local and state outlets, but the Canadian Broadcasting Corporation, our friends in the north, had also found us. So had National Public Radio's *All Things Considered*. After many months of national media inattention for Flint, I didn't want to miss any opportunity to be heard. I didn't want to grow tired of explaining—or become overwhelmed.

To make things easier, I told reporters to just call me what my patients did: "Dr. Mona." Otherwise, I've got a long and unwieldy string of names, too foreign for some to pronounce. And when I'm Mona, the name Haji presciently gave me, or even Dr. Mona, I feel more relaxed. It seems friendlier and more personal. I felt that way about the media—and really grateful. It was the day-in, day-out work of local reporters Curt Guyette and Ron Fonger, in print and online, along with Lindsey Smith from Michigan Radio, and others, that kept the Flint story alive.

Wurfel stayed on message and continued to try to discredit me. With the release of my research, the water controversy in Flint was reaching "near-hysteria," he said. There it was again, the sexist *hysteria,* the word used against Alice Hamilton. I wore the insult like a badge of honor.

Most mornings, usually by 7 A.M., I had a conference call with Dr. Reynolds, Kirk, and Jamie from the United Way to go over the day's developments and plans. Sometimes Andy and Senator Ananich joined in. We were incredulous at the state's all-out resistance. We were horrified, on fire. Once I started swearing, I really took off. Andy threw wicked insults at our antagonists—*dumb bastards*—but Marc Edwards produced the most ingenious ones. To him, Wurfel wasn't just dumb or a bastard. According to Marc, he was a "fucktard," which, according to Urban Dictionary, means "an extraordinarily stupid person who causes harm." I had never heard the word before and didn't like it, to be honest, but it worked.

Meanwhile, my wish list for Flint kids kept getting longer. School health. Early literacy programs. Nutrition. Transportation services.

Nobody on the team grew disenchanted. We talked through our options, discussed next steps, and strategized about how to keep pushing the state to deal with the water, which it was still refusing to do. We wanted filters and bottled water brought to the neighborhoods with the worst water first. We wanted them home-delivered. *Where is the National Guard? How do they get activated?*

We all had our areas of expertise, but it was almost comical what we *didn't* know, which was almost everything. Some days we met in person—or patched in the others—and wondered aloud, *How do emergencies usually get handled?* It was the blind leading the blind. If we had been dealing with a chemical spill or a flood, the proper authorities would have been "activated" and ordered to help long ago. When Toledo had had a water issue less than a year before, water trucks arrived in less than twenty-four hours. Where was the sense of urgency here? We wondered why the state didn't send support, but then we realized that sending in relief at such a large scale would be an admission of blame.

So the stalling continued. And without a command center or authority of any kind, first-responder responsibilities fell to the local United Way. Our goal stayed the same. We needed to get people filters and bottled water. And after that, we had to explain how to install the faucet filters. Jamie was working with the food bank and the Red Cross, figuring out how to get donations and coordinate involvement from big community groups, and working with residents and volunteers to get it all done on the ground level. He was tireless in taking on new tasks.

Ahead, we had a million steps to take before we got to the impossible goal—getting the water source switched back to Detroit. But we never lost sight of it. We stayed focused on deeds that led to action, on words—at least in public—that helped rather than hurt, on a perspective that handled triage and short-term obstacles but never lost sight of where we wanted to wind up. We were always in contact, connected, thinking together. At least once a day, Marc and I caught up—and quickly became a tag team. He was the master of water-safety issues. I took on all health matters and anything that involved Flint kids. The pushback from the state just gave us more fight.

As soon as there was a new development, which was several

times a day, I updated the rest of my team—Jenny and Elin—and heard their ideas. The rest of the minutes of the day were an obstacle course of calls, responses, interviews, and disappointments. It was an all-consuming campaign.

I relished each small sign of progress. And as many times as the hurtful words "splice and dice" emerged unbidden into my mind, I also hung on to the moments that gave me hope. It wasn't until a break came that I realized how much I needed one.

—

CHAPTER 22

Demonstration of Proof

A BREAK CAME THE WEEK FOLLOWING THE PRESS CONFERENCE. Out of the blue, I received a call from Eden Wells, the chief medical officer for the state of Michigan, whom I had met over the summer. She apologized for "not reaching out earlier." She wanted to have a "physician to physician" conversation.

I was surprised to hear from her. I was also excited, gratified, hopeful, and skeptical, the wild stew of conflicting feelings I was getting used to by then, something that came with almost every development. First there was paranoia: *Why does Eden Wells want to get in touch with me?* A few minutes later, a sense of optimism and relief. *Wow, the chief medical officer for the state of Michigan! This is great news. Finally, an opportunity to cut through the crap.* And then skepticism: *Nothing will happen. Again.*

On the phone, Eden didn't sound like a monster. Her voice was normal and friendly. She remembered our meeting at the vaccine press conference in July, which now seemed like years ago, and where we'd shared a laugh. My white coat said Michigan State; she was on the faculty at the University of Michigan. That was supposed to mean we represented two rival schools, except I don't care about rivalries, or football, and besides, I had attended and love both universities.

She explained why she was calling: she was now the state's "point person" for the water "controversy." I had several conflicting reactions to that news but kept them to myself. She wanted to ask a few questions about our research—and how I'd arrived at the findings.

So I told her about our methodology, referenced past research done in this area, and explained how Jenny and I had reached our decisions about how to do the study—and why our numbers were different from the state's.

We were running our numbers again, I said. This time we were using GIS software to be more precise. Then I mentioned I was still waiting for the state's blood-lead-level data, the entire spreadsheet of each test for every child, not the elementary one-page tally sheet they had released earlier. Our request was still pending. By then, we had even submitted an official IRB application to the state for their raw data.

Eden said the state was going to relook at its data—so she could compare "apples to apples"—and that she would try to expedite my request for the state's raw data. The way she sounded, it seemed like it wouldn't be a problem. And I found myself feeling relieved—even good—for the first time in a while. I trusted Eden on some level. I even offered to do my own GIS-based analysis of the state's data and to give Eden the results in a day, in a desire for complete transparency. The state office didn't have GIS experts.

Finally, I was having a peer-to-peer conversation with another scientist and doctor, discussing my findings and how I got there.

Finally, somebody was thinking about the science and methods instead of ridiculing me and calling me "unfortunate" and "hysterical."

Following up on our call, I sent Eden published articles about how lead in water spikes in the summer and how I controlled for it. I shared recent research from Montreal scientists showing how every 1 ppb increase in lead in water increased blood-lead levels 35 percent. And I told her how GM had stopped using this water because it was corroding engine parts. Privately, I wondered how much the *Detroit Free Press* story, which called the state's numbers into question, had provoked her call—and got the governor's people to withdraw from the attacks and actually look at their data and my research.

Later on, when the Freedom of Information Act requests went through and the agency emails became public, some of my questions were answered. It turned out that when Eden called me, she had already received an email from her boss at the MDHHS, Nick Lyon, directing subordinates to prove that I was wrong, even before they'd seen my study, even before they'd looked more closely at their own numbers. Lyon's office had been commissioned to look into the water-quality complaints months before. And when they saw a spike in blood-lead levels—as Karen Lishinski had revealed—they had been blinded by preconceptions and reached the conclusion there was nothing to worry about. Now, in the face of my research, Lyon asked his department to "make a strong statement with a demonstration of proof that the blood-lead levels seen are not out of the ordinary." In other words: he was telling the public servants to stand firm that their results were right, whether they were or not. Don't bother rechecking them. Instead, go on the attack.

The state determined that my Achilles heel had to be seasonality. It was the only remotely credible argument that could be used to discredit my research.

Didn't Lyon care about protecting kids?

Didn't he care about science?

No, I think all he cared about was winning. It was risky, a gamble. He had to count on the fact that nobody else in his agency, or the governor's office, would come along and care about kids or science either. If folks could claim global warming is a hoax, tobacco is good for you, and asbestos is a safe building material, the Flint blood-lead levels should be easy enough to obscure.

But I think Eden Wells cared about science, or appeared to. That's what caused the tide to slowly turn, what eventually stopped the cover-up, the lies, and the stonewalling.

When Lyon sent out his directive to make me look bad, Eden sent out an email to colleagues with a caution:

Dr. Hanna-Attisha is doing a GIS Analysis now and that is why she really wanted the data. She has a group of Epis and statisticians behind her. She has a strong foundation in research. She would understand the limitations of the data.

Was she sticking up for me? I'll never know. Maybe she decided that it wasn't so smart to attack and discredit somebody with real research chops (although there was no group of epidemiologists and statisticians behind me, just Jenny). Or maybe the media stories were beginning to get traction, and she feared that her own reputation as a public health professor was at risk. But I choose to believe that she really cared about good science and about kids.

Either way, whatever the truth—or whatever combination of pressures and factors was at play—I think if it weren't for Eden, the state's denials could have continued months longer, maybe years. Flint might have been another D.C., with agency officials digging in their heels year after year, until nobody cared anymore and the pipeline to Lake Huron was finished anyway. But instead of pushing back at me, Eden made her scientists rerun their numbers, using correct methodology.

On October 1 the Genesee County Health Department declared a public health emergency, and the governor's office announced it was having a news conference with Nick Lyon the next day at 1:30 P.M.

I heard from Andy immediately. The governor's office and Lyon were going to confirm my findings tomorrow. The state had surrendered.

Holy shit. You have got to be kidding me. It was hard to believe or process.

But Eden called and confirmed it, on her way to the press conference. "We ran the numbers as you did, and the state's blood-lead numbers *are* consistent with yours," she said.

I thanked her, kind of numbly, not sure how grateful I was supposed to be.

"And at one-thirty," she went on, "you will get access to your state data request."

All the blood-lead levels for all the kids in Genesee County. *Finally.* But it was weird that it was being released at the exact same time as the press conference. Did they think I would leak the info if I got it ten minutes earlier? The state was changing its tune, but its lack of finesse and strange ways continued to bewilder me.

TECHNICALLY, I WASN'T INVITED to the official state news conference, but I wanted to go and hear what the state had to say, directly, for myself. That way I could gauge the reaction and get a feeling for what might come down next. If Kirk and I had a chance to corner somebody in power, even better. We still needed an emergency responder. And we wanted to continue our drumbeat about getting the water changed back to Detroit.

Kirk figured out a way to get us both included. He picked me up at Hurley in his silver Jeep and drove to a small, out-of-the-way building, the Kettering Innovation Center on the Kettering University campus. Kettering, who had pulled so much lead out of the earth and poisoned so many kids.

There were protesters outside. We were happy to see them, and I thanked them for being there, for their activism. I wanted to protest too.

Inside, the small room was packed with media. Kirk and I separated and were squished into two different sides of the small room. At the front, Eden Wells, Nick Lyon, and Mayor Walling were hovering around a podium and preparing to talk. I looked around and noticed Dan Wyant, the head of MDEQ—the guy in charge of the Drinking Water and Municipal Assistance Division, in charge of all those people in white lab coats who'd been entrusted with making sure our water was safe.

Brad Wurfel was in front, facilitating, or appearing to. He was tall and cartoon-handsome, with a pearly-white smile—a classic public relations guy. It was part of his job to look chill, but I sensed that under the surface he was sweating and scared. Or he should have been.

Eden opened the conference by saying the state had looked at its data again, and although the numbers weren't exactly the same, there was an increase in the percentage of kids with elevated blood-lead levels.

"We understand many have lost confidence in the drinking water," Dan Wyant said to the room. "We need to build that back. We need to do that more."

There was an unveiling of a ten-point "comprehensive action" plan. It was not comprehensive. It was not action. It was basically a logo. It did include testing the drinking water at Flint schools, but otherwise it was pulled together at the eleventh hour by people with no experience, depth, or foresight and definitely no public health experience. Thank God for the media, because reporters were asking great questions:

"Is corrosion control being used?"

"Yes," Wyant said. Unwittingly or not, he was wrong.

"Do you think the Genesee County Health Department should have declared a public health emergency yesterday?"

"No," Lyon responded.

What the fuck? I whispered under my breath. But there was more dumb stuff to come. Finally the state had stopped fighting my research and conceded there was a problem, but it still seemed to be fighting the idea that it had anything to do with the creation or mitigation of the disaster.

After the press conference, Kirk and I went into a different room and had a chance to talk to state officials, to dig into the details of what seemed a superficial ten-point plan and to insist that the water source be switched back to Detroit. I had a chance to meet feisty Sheldon Neeley, who represents Flint in the state legislature. He was mad at the state and took a break from a few heated exchanges to thank me for my work. I gave him a hug. Later, I heard a rumor that he showed up in Mark Valacak's office with a baseball bat, demanding that a public health emergency be declared.

Across the way, I saw Eden Wells and greeted her. I was excited and eager to check out the state's raw data, a larger sample, and do my own analysis on it—our holy grail.

Eden was cordial. So was I. But not far away, a shouting match exploded between Jamie and Harvey Hollins, the director of urban initiatives, who seemed to be Governor Snyder's new point person for Flint. Although the United Way was part of the governor's new plan ("call 211 for filters and water"), nobody had contacted Jamie yet, and he was furious. The fight was face-to-face and tense. Hollins seemed rattled and hadn't expected a confrontation. Only the press were supposed to be there.

As I took this all in, feeling both horrified and gratified, suddenly I realized that Brad Wurfel was right next to me. He was towering over me, like a human embodiment of the state's ten-point plan: all suit and hair, no substance.

I couldn't help myself, because I believe in standing up to bullies.

"You called me unfortunate," I said, looking up at him. "You said I was irresponsible. You said—"

He grimaced, then his face turned very solemn. "I'm sorry," he said.

Wow, that surprised me. The monster actually seemed to feel remorse. *Fucktard said he was sorry.* But my next reaction was unexpected. I thought, *Why the hell is he grimacing and solemn and apologizing to me?*

I wasn't the person he should be apologizing to. Every kid in Flint deserved one. Every resident of Flint. They all deserved an apology and much more. Wurfel's was the first of many I'd hear the next year, and every single one made me feel awful.

I nodded. *Okay. Fine. Get it off your chest.* I had a hunch he wouldn't have his job much longer, and I was right.

Then Dan Wyant approached me—the head of MDEQ. He mumbled some kind of greeting and explained that he "meant" to make a public apology to me, but the opportunity hadn't come yet.

He sifted through some papers in his hands, his prepared remarks and talking points, desperately looking for the piece of paper where his apology was written out.

"If someone asks, this is what I am going to say," he said.

I felt sorry for him, to be perfectly honest. He was probably a nice guy, seemed like he meant well. But he was in over his head.

Forget the nonapology and the mumbling modesty. I had something I had to ask. "Why wasn't corrosion control added?"

He gave me the same blah-blah-say-nothing response that we'd all been hearing on the radio and TV for weeks. It had finally been discredited, but he was still saying it. He didn't have his job much longer either.

The next day, Saturday, October 3, water filters were distributed to the public for the first time, at the University of Michigan–Flint campus. Marc Edwards advised us on the brands that did a good job filtering out lead. NSF 53–certified lead-clearing filters were purchased and provided by the United Way, thanks to Jamie's efforts and thanks to online fundraising by Marc's engineering students. I stayed home to finally spend some time with Nina and Layla, but my residents and medical students volunteered at the filter distribution and kept me informed about how it was going.

Filters were prioritized for pregnant moms, formula-fed babies, and zip codes where the water was known to be the most toxic.

We worried about long wait lines. We worried about running out. We even worried about riots. Flint residents were rightfully angry.

But the lines moved quickly, peacefully. The filters were distributed. It was happening, finally happening. At the time, I thought it might be the most we could hope for.

I sent Jenny a text.

ME: Jenny, we did it. Kids are going to be protected.

—

CHAPTER 23

All the Things We
Found Out Later

S O MANY OF THE THINGS WE FOUND OUT LATER WERE TROU-
bling.

Exchanges between state officials in 2015, released through
FOIA, many of them Marc's requests, show how they defended
their flawed science. In shockingly flip and derisive language, they
spent so much time creating extensive "talking points" to explain
why they were right and everybody else was wrong. They made no
attempts to get to the bottom of anything; their only goal was to
recast and massage their lies.

We found out later that in January 2015 state officials, while telling Flint residents that their water was safe to drink, were arranging
for water coolers to be delivered to the Flint State Office Building
so state employees wouldn't have to drink from the tap.

We found out later that as early as March 2015, the governor's
office was exploring the distribution of water filters in Flint. Later
it finally worked with two companies to donate to the Concerned
Pastors for Social Action, which began handing out filters on September 1. At that time the state was still officially denying the water
problems to the media, to outraged citizens, and to me.

We found out later that the city, controlled by the state, had deliberately manipulated the water samples from Flint homes so they wouldn't have to notify the public about the presence of lead, per the federal rules. To ensure a low percentage, they collected samples in such a way that caused less lead to be detected: they pre-flushed, removed faucet aerators, used smaller bottle mouths, and tested in homes without lead service lines or lead pipes. When high-lead samples came back, they were thrown out, including those from LeeAnne Walters's home.

We found out later that after Miguel Del Toral's testing at Lee-Anne's home in April 2015 found hazardous-waste levels of lead in the water—and after he raised the alarm to his agency, the EPA, and MDEQ—he was dismissed and discredited and criticized for involving himself in matters that didn't concern him.

The EPA would eventually conclude that MDEQ should have supervised optimized corrosion control as soon as Flint switched to Flint River water. This was the critical failure. But instead MDEQ, once alerted, hunkered down to dismiss the problem and lie about it. The drinking water office at MDEQ seemed determined to approve permits above all else, not to evaluate the science and public health implications.

No one really knows why corrosion control wasn't added to the water—it would have cost only eighty dollars a day. Or why the pump to deliver corrosion control was never installed, or why so many other shortcuts were made and so many people were ignored who tried to raise red flags.

But we do know that in July 2015, an MDHHS analysis showed that blood-lead levels had spiked in the summer of 2014. Having missed it for a whole year, the MDHHS had an analysis done—with real science and fancy stats and fancy graphs—and then promptly dismissed and covered it up. This is the spike Karen Lishinski had inadvertently revealed to me—the results that were never shared with me when I asked for them.

When our work challenged the department to take action, Wesley Priem, a manager of the MDHHS Healthy Homes Section, responded by writing to a department colleague: "Yes, the issue is moving . . . at the speed of rushing water. . . . This is being driven by a little science and a lot of politics."

We found out later from the emails that as early as December 2014, red flags were being raised about a strange escalation in cases of Legionnaires' disease, a severe, often lethal lung infection caused by waterborne bacteria sometimes found in water towers of hospitals, hotels, and large institutions. The number of Legionnaires' cases quadrupled after the water switch—also related to the lack of corrosion control—yet nothing was done. Two top staff in the governor's office were notified as early as March 2015 and recommended action, yet no action was taken and the issue didn't become public for another nine months. The county health department, which also knew about the increase in cases, did not bother to alert providers or the public to be on the watch for Legionnaires'. This led to eighty-seven cases in 2014 and 2015, at least twelve deaths, and an uptick in pneumonia mortality.

Again and again, the state and federal officials' disdain for Flint was shocking.

At the EPA, when asked about using federal money to buy water filters for city residents, the Region 5 Water Division chief, Debbie Baltazar, wrote to the regional administrator and others, "I'm not so sure Flint is the community we want to go out on a limb for."

The pointed cruelty. The arrogance and inhumanity.

Sometimes it is called racism. Sometimes it is called callousness. And sometimes—when the Legionnaires' disease outbreak that left at least twelve people dead was tied to the water switch, something the bureaucrats knew about for a full year—it can be called manslaughter.

THE NEXT BATTLE WAS the "impossible" one: getting the water source switched back to Detroit. Every time we raised it, we were totally shut down. Everybody said it was impossible. Never going to happen, never. It wasn't just about money, they said; it was about engineering and ownership of the pipes, legal liability, and *forget it, never going to happen*. But we weren't going to give up. They had lied about so much already. They could be lying about the water switch too.

So we hunkered down for a long fight.

Distributing filters and bottled water was not enough. Government had to provide a long-term solution. People cannot live on filters and bottled water forever.

The excuses we were given—that the pipe was already sold, that it was too expensive—turned out to be paper tigers.

The media grew more vigilant and critical—specifically the *Detroit Free Press*, which in October 2015 called the water crisis an "obscene failure of government." Flint residents and activists said far worse and were finally heard. Everybody now knew there was something wrong with the water—and specifically *what* was wrong. Flint residents got even more angry, organized, and mobilized, especially since the harm was focused on young children.

The activists held bigger and bigger demonstrations, where Jesse Jackson and Michael Moore made appearances. Residents channeled their anger into bottled water distribution and into Karen Weaver's campaign for mayor; the election was just a couple of weeks away, on November 3.

The Charles Stewart Mott Foundation's new president, Ridgway White, knew me—we had worked together for the past year building our new clinic atop the Flint Farmers' Market. As soon as he saw the findings of my research, he called the governor and pushed him to switch Flint's water source back to Detroit water. He even offered the assistance of the foundation, a private grant-making organization with $2.7 billion in assets.

Shortly afterward Natasha Henderson and Howard Croft hosted a heated technical advisory committee meeting, where two good scientists from the EPA, Darren Lytle and Mike Schock, colleagues of Miguel's, recommended the switch back. But the decision had already been made.

I was with Allison in the clinic on October 8 when Governor

Snyder announced that the water would be switched back to Detroit. Allison was seeing patients, coming and going between exams. I was glued to the live event on a computer monitor. The "impossible" water switch was actually going to happen, maybe within a week, the governor said, and it would be accomplished with a combination of funds—city, state, and a generous $4 million donation from the Mott Foundation.

Andy wrote to me and Marc: "There is no way that this ever would have happened without the two of you, so thank you for all the lost sleep and incredible work you have done on behalf of Flint residents."

Melany wrote, "You should get a cape!"

Marc wrote, "You are the woman of the hour."

I was numb, watching in foggy disbelief. *Is this really happening?* Exactly two weeks before, I had stood in the Hurley conference room, presented my study, then curled up in a ball on my bed while the governor and his henchmen made what seemed like a coordinated effort to destroy me. Now there he was, the head of our state, speaking in his businesslike way about the water switch and taking way too long to say it. His words blew over me like a rush of hot air. I tried to hear what he said and believe he cared. I tried to absorb the victory and feel rewarded. But there was nothing victorious about it.

During the same press conference, another announcement followed: toxic levels of lead had been found in the water in three Flint schools. Immediately afterward, Eden Wells came forward and spoke for what felt like thirty minutes about all the usual sources of lead—paint, paint dust, and so on—and the need to vacuum and mop and clean windowsills. Talk about a confusing message.

I couldn't believe it. Right on the heels of admitting that there were toxic levels of lead in the school water, she was talking about vacuuming. I was furious. I thought we had been making progress. I thought we were past blaming the victims. I thought we were teaching people about lead in water. *Why is she going on about paint?*

Kirk texted me immediately.

KIRK: 3 elementary schools, 1 over 100 ppb

ME: Holy shit

KIRK: watching live. Heartbreaking.

ME: I'm watching now. I hate them.

I should have been happy about the water switch, but how could I be? *Three schools with toxic levels of lead.* I walked to my clinic office, closed the door, and tried to keep myself together. My hands were shaking, my whole body was shaking.

That same day, I was hosting a group of pediatricians who had arrived in Flint for an AAP grant site visit. Without a chance to reflect, I dried my eyes and tried to pull myself together.

And it was almost time for my resident conferences.

My residents, all twenty-one of them, were another group besides my immediate family who suffered from my absence. I didn't know this year's first-year residents as well as I usually did by now. And the second- and third-year residents probably saw the biggest change, having known both pre- and post-Flint-crisis Dr. Mona. Since the end of August, I just hadn't been around or involved as much as in the past. I told myself now, finally, with the water about to be switched back, I had to explain, apologize, and make up for losses.

Earlier in the week, I even sent them an impassioned group email:

FROM: Mona Hanna-Attisha
TO: Pediatrics Residents; Combined Internal
Medicine/Pediatrics Residents
SUBJECT: This is our fight song!

All,

It's been a crazy month, and I wanted to apologize to all for being preoccupied. I'm so sad I missed the Peds retreat yesterday! As you all may know, I have been consumed by the lead in water issue. Many of you have been at the front lines of this issue, especially the Community Peds residents.

This has been an incredible example of our power and credibility as pediatricians. When the pediatricians spoke (armed with data—yeah, research!), we changed the game and inevitably, the future of Flint's kids.

"Never doubt that a small group of thoughtful, committed citizens can change the world. Indeed, it is the only thing that ever has." Margaret Mead

This story is not done and our research/activism continues. We are now doing really cool geospatial analysis thanks to MSU public health research superstar Prof Sadler, we are busily writing a manuscript for publication (thanks to Allison), and we have a whole list of subsequent research projects planned. If anyone is interested in participating in anything, let me know!

This is our fight song!

Mona

LATER ON, MANY CELEBRITIES, athletes, and politicians would stop in Flint to show support. We saw President Obama. We saw Hillary Clinton and Bernie Sanders, who had a presidential primary debate in Flint. Even an orange-haired Republican candidate for president turned up later.

But the first visit was from Senator Debbie Stabenow, who came to Hurley on October 16, the day the water was switched back. Her visit was, in many ways, the most important.

I had been wanting to meet Senator Stabenow for years. My brother, when still an undergrad, had worked on her first campaign for Congress. And Elliott worked closely with her on children's health initiatives, related to his mobile health work in Detroit schools. Senator Stabenow, whose mother was a nurse, understands public health issues and understands lead and its pervasiveness. And as a former social worker, she has always been a strong advocate for children.

Melany and I spent an hour with the senator, sharing my research. She was very upset about what the lead exposure meant for kids, but also at how disenfranchised Flint was, its fate in the hands of an unelected emergency manager and an indifferent state. I summarized the issue in a practical one-page takeaway sheet—something all good advocates learn to do—that described potential next steps and specific things she could do to help on the federal level.

Senator Stabenow got visibly angry and wanted the governor to make a "disaster declaration," which would free up more state and federal resources. She also supported a full investigation, forcing the state to take full responsibility. But that alone wouldn't be enough. The Flint water crisis was a national emergency, she knew, requiring the nation's collective help. She promised to fight for federal funding for long-term support.

All these steps were critical but also far off—months or more down the road. I wanted something right now, something that would make a big difference immediately. I knew Senator Stabenow sat on the Senate Agriculture Committee and was a big supporter of WIC and expanded nutrition programming.

So I made my request, for Nakala and all the other babies in Flint who currently subsisted only on formula mixed with Flint water: "We need your help to get ready-to-feed formula for our babies." It would require a federal waiver from the USDA. She promised to put the weight of her office on making it happen. When we said goodbye, I had to give her a hug.

Many politicians came to Flint after that, and I made a point of giving them all hugs. Somebody started calling it "Hug Diplomacy."

That same day, October 16, the EPA reversed course and acknowledged the water crisis. It established the Flint Safe Drinking Water Task Force—with Miguel on it—to help develop and implement a plan for secure water quality in Flint. It eventually issued a

WITH SENATOR STABENOW, OCTOBER 16, 2015

directive to prevent state water agencies from fraudulent testing methods that purposely minimized lead content.

I COULDN'T HELP BUT think of all the excuses we had been given for why the water couldn't be switched back.

All those reasons, explanations . . .

"The pipe has already been sold."

"It's prohibitively expensive."

"Impossible!"

It happened at 5 P.M. on October 16.

With a simple flip of a switch, noncorrosive water once again flowed into the taps of Flint homes.

CHAPTER 24

Fire Ant

A S SOON AS THE WATER WAS SWITCHED BACK, OUR DRUMBEAT shifted to demanding that a state of emergency be declared. The switch alone wasn't enough to deliver clean, safe water to the city. Eighteen months of corrosive water had done great damage to the underground network of water pipes, and it had also corroded the pipes and appliances in residents' homes. The water was not yet safe to drink. Much more remediation was still needed.

I dove headfirst into two all-consuming projects. One was perfecting my list of science-based demands and recommendations for Flint kids. It had grown long and unwieldy and needed to be focused. The other project was producing the final, unassailable, and more accurate publication of elevated blood-lead levels after the water switch, using geospatial software.

The rules of academic research didn't allow my original study to be peer-reviewed and published in a journal because it had already been presented and made public. But a new study, using GIS, could be. An article with a scholarly stamp of approval would make the crisis real in terms of science. And once it was published in an academic journal, it would be cited and referenced, much harder for anyone to shoot down.

Rapidly, over the course of the following week, Jenny and I doubled our team of researchers. Rick Sadler, working with GIS software, created incredible maps, even better than the previous ones, that presented unforgettable images of the impact of corrosive water on the blood-lead levels of Flint children. My resident Allison, now our appointed lead water researcher, spent every free moment reading the available literature, swamping our medical librarian Sharon Williams with requests for articles. Jenny and I began writing while continuing to dig, working with new numbers for the specific neighborhoods where the water tested highest for lead.

Marc Edwards, between his academic obligations at Virginia Tech and his research, never stayed in one place very long. Even after he left Flint, he stayed in touch and guided us when we needed him. He made another contribution too: infectious paranoia. His stories about working on the D.C. crisis were disheartening and scary. He told us about an incident when study figures had been stolen from his lab, and another when a member of his research team was discovered to be an industry mole. Because of what Marc had gone through with federal agencies—they dismissed his work for years and tried to deny that lead in water could harm children—I decided not to reach out to the CDC, worried that their response would be impossible to predict.

Marc had hacking stories too and warned us against communicating in email and texts. Partly worried and partly amused, Jenny and I kept our new study on an encrypted USB only, and drove it back and forth between our offices at Hurley and Rick's office at MSU.

We used code names for each other in emails too, to protect our identities—and to allow us to be cynical and use expletives liberally. I have to confess that I found this kind of exciting; it reminded me of the *noms de guerre* of my uncle Nuri—"Anwar"—and many other dissidents at the time. (Even Haji had a code name, "Faris.") Jenny was Red Panda. Allison was Little Bird. Rick was DikDik, which had something to do with his summer camp nickname as a boy. I was Fire Ant. I liked that.

Working quickly, with the tight, all-consuming focus that was becoming habitual for us, we finished the paper by late October and gave it to Elin, Marc, and Nigel Paneth, a pediatrician, epidemiologist, and distinguished professor at MSU, to review. They all gave us critical feedback and suggested revisions. Then we submitted it to the *Journal of the American Medical Association,* one of the most prestigious medical publications around, but it responded that it offered a rapid review process only for drug trials. Jenny quickly found that *The American Journal of Public Health* was willing to peer-review it as soon as possible. "Elevated Blood Lead Levels in Children Associated with the Flint Drinking Water Crisis: A Spatial Analysis of Risk and Public Health Response," coauthored by all four of us, was accepted for publication on November 21 and went online a month later, right before Christmas.

Having the study peer-reviewed and published protected me from further criticism and laid the groundwork for more serious recognition of the crisis. It also gave me leverage when time came to talk about long-term care and prevention.

BUT MEANWHILE THE STATE was still playing catch-up and making some very stupid moves.

For starters, the governor's office still seemed to downplay the health risk of lead to children, even after eighteen months of exposure. I guessed the staff was trying to make the crisis, and therefore their own complicity, seem less awful. And of course, money was a factor. The state had no budget yet to make a real response, and the scope of the liability for the harm caused was a big unknown. Lawsuits were being filed daily, a mix of class-action suits and individual cases, some by lawyers for the ACLU and the Natural Resources Defense Council on behalf of Flint residents. Some sought damages. Some sought immediate relief, like enforceable commitments to replace the lead pipes and to provide services for kids.

The intense panic in the governor's office and the two agencies involved, the MDHHS and MDEQ, became obvious at a press conference on January 11, 2016. The plan was for Governor Snyder to come to Flint and meet with Karen Weaver, the new mayor, for the first time. (Dayne Walling had lost his bid for reelection on November 3, one of many political casualties of the water crisis.) Hoping to marshal support, the governor's office called and asked me to attend.

I was conflicted about going. I didn't want to be used—or become a Flint water crisis trophy for the same state government that had ignored the complaints of residents all these months and that would have continued to ignore them in perpetuity. And besides, I didn't really know what the press conference was about. At the same time, I didn't want to miss an opportunity to advocate for Flint kids. When I called Andy for his opinion, he said he didn't know what it was about either, but he and Senator Ananich planned to be in the audience.

The initial meeting place was in the basement of Flint City Hall, where cabinet members and the new mayor gathered before the press conference. Eden Wells saw me and said she wanted to share something with me—preliminary graphs of the deaths from Legionnaires'. A quick glance said it all—there had been a huge spike in Legionnaires' after the water switch. While I was processing this terrible news, which I hadn't known much about previously, Nick Lyon appeared, glanced at the graphs, and said, "Can't we just say this is due to seasonality too?"

I raised my eyebrows. *Is he serious?* Instantly I had a really bad feeling. I was in a room full of the people who had poisoned a city to save money. I was behind enemy lines. What was I doing with these people, and *what was coming next?*

Then the governor arrived. He had just left a heated meeting with the Flint pastors and something traumatic must have occurred there, because he was visibly shaking. I had never met the guy before, even though he'd been planted in my brain for weeks now. We greeted, exchanged niceties, *Thank you for coming. Thank you for inviting me. . . .*

On our way to the press conference, in another room of City Hall, I could hear protesters chanting both outside and inside the building. I would have given anything to join them. But instead, I paraded in with the rest of the suits and landed in the second row directly behind the podium. A sea of reporters, microphones, and TV cameras spread out before us, but weirdly, the only people sitting in the auditorium were Senator Ananich and Andy.

I was a trophy—and captive.

AT FLINT CITY HALL, DURING THE GOVERNOR'S PRESS CONFERENCE, JANUARY 11, 2016, WITH EDEN WELLS (FAR LEFT), GOVERNOR SNYDER (AT PODIUM), KAREN WEAVER, AND NICK LYON (AT RIGHT)

The onslaught of misinformation and lies began. First, MDEQ blamed the high water-lead levels in the schools on the school fixtures. Then Nick Lyon stepped up to the podium and stated unequivocally that since October 1, 2015, forty-three kids had been exposed to lead.

He had a number.

It was forty-three.

I couldn't believe what I was hearing. Had he really said forty-three? How had he arrived at that wholly inadequate number? Who had given it to him?

It was outrageous—medically ridiculous—to announce that only a few dozen kids had been exposed. The time frame was way, way off. By October 1, 2015, the exposure had been going on for eighteen months—and it was two weeks after my big press conference, when the word had already gone out to parents not to drink the water.

What about all the kids who had had elevated lead levels before October 1?

Another thing: only kids on public assistance had mandatory screenings for lead exposure, and less than half of those kids even got them. *What about all the others?* There were thousands of kids in Flint under six. And on top of that, the number didn't account for pregnant women, newborns, or infants who drank the Flint water while MDEQ authorities were declaring how safe it was. Babies and children under one were not even tested, and when they were, it was well after the exposure was realized, and often too late for the lead to show up in their blood.

Every kid had been exposed.

Or, 8,657 had, to be exact, according to the census.

The state said it wanted to earn back the residents' trust and make amends. But it didn't want to admit the obvious: that the entire city of Flint and all its kids had been exposed to lead in the water—whether it was at home, at school, at Grandma's house, or in a neighbor's kitchen. Lead was in the water that had been mixed with formula and powdered lemonade and other drinks. Lead was in the water that had been used to boil macaroni and vegetables, to cook instant ramen. And when a child turned on the water fountain at a Flint school, lead had been in that water too.

I don't have a poker face. This is why I can win at Konkan, because it isn't a game of bluffing. But this time, my lack of reserve worked for me. In a matter of seconds after Nick Lyon opened his mouth, a dozen TV cameras could see that I wasn't on board: I was shaking my head, shaking and shaking it. And every time Lyon said another stupid thing, I shook it again, visibly and actively and forcefully protesting. The governor's office had invited me, but that didn't mean I had to toe their line and smile like a fool. I caught a glimpse of Andy and Senator Ananich in the audience. They seemed despondent, kept looking at the ground. They were just as angry as I was.

As soon as the press conference ended, I shot an email to the governor's office explaining that yes, the fixtures in the Flint school had lead, but it was released when they came in contact with untreated corrosive water. And saying that only forty-three kids in Flint had been exposed to lead was bogus. Next, I contacted the media, protesting the state's weak understanding of population-wide exposure. To properly address the public health crisis in Flint, every child under six years old—8,657 kids—should be considered exposed to lead.

I worried that if I didn't protest loudly and persistently, the downplaying and misinformation would live on. And indeed it did. The number forty-three was picked up again and again, even cited by *The New York Times,* to discuss the extent of the problem. But worst, in my mind, was the Kettering University president who used "forty-three" in a letter to students, staff, families, and alumni in an effort to keep Flint's water from adversely impacting enrollment. When I heard about his letter, I contacted the president, who quickly posted a correction, removing the blood-lead data. But he stayed in damage control mode. Kettering was again living up to its namesake.

Something else lived on and on: the clip of my head-shaking protest during the governor's press conference. It was on the nightly news. It made the rounds on the Internet. And probably because Rachel Maddow has an incredible research team that doesn't miss anything, the head-shaking clip found its way to her show, which was way out ahead on Flint with cogent analytical segments, a mix of anger and concern for the kids.

She smartly connected the crisis to the undemocratic EM law, which she labeled "radical change" from "the way we govern ourselves as Americans." And she was the first to say that the kids of Flint had been "poisoned by a policy decision." As for the head-shaking clip, she said I was "badass." I can't imagine a better compliment.

Within a week, the state finally capitulated, and I finally felt we were on the same page, using the same numbers and the same words. The turning point came when Eden Wells told the *Detroit Free Press* that lead exposure was population-wide in Flint and not limited to a few dozen children. From a public health perspective, she said it was important to "consider the whole cohort" who had been "exposed to drinking water"—especially kids six and under who had been consuming the water since April 2014. The number forty-three vanished as mysteriously as it had arrived. All the kids in Flint had to be regarded "as exposed," according to Eden, "regardless of what their blood level is [today]."

That was a victory. The future for the kids in Flint came down to a few coded words: "whole cohort" and "regardless." Without a shared understanding of the problem—that all children needed to be protected and treated, whether or not they had problems, or will ever have problems—we couldn't hope to arrive at solutions that would help everyone.

SHORTLY AFTER TAKING OFFICE, Flint's new mayor, Karen Weaver, declared a citywide state of emergency. She had no formal power to do it, but around a small room, many of us urged her on, including a newly hired emergency responder for the county—and it worked. The declaration had a ripple effect and caused a new flood of media attention, forcing Governor Snyder to declare an emergency in Genesee County a few weeks later, which got the National Guard on the ground to distribute bottled water and filters, and which prompted President Obama to declare a "federal" emergency in Flint on January 16, 2016.

Being declared a "disaster" would have been better—and freed up even more resources—but it turned out that would have required a natural disaster like a fire, flood, or earthquake. It made me mad, since the potential consequences of the lead exposure are far longer lasting. Nevertheless, the federal emergency declaration was a big deal. For the most part, the feds knew what they were doing. It was no longer amateur hour. They understood the scope of the impact, and I never had to explain what toxic stress was. They got it, and they tried. Dr. Nicole Lurie, a seasoned physician and a uniformed U.S. public health service officer, took charge and coordinated the response from every acronym agency imaginable—FEMA, HHS, SBA, EPA, CDC, HUD, SAMHSA, and more.

By then, Jenny, Kirk, and I had been working on a list of interventions for a couple of months. I had looked at studies, researched programs, and called national lead experts, child development experts, and toxic stress and resilience experts. We wanted to recommend programs that had been tested and were proven to work. We let science tell us the best ways to protect children, promote their development, and improve their outcomes. This is what kids need everywhere, but especially Flint kids.

Most important, we needed to get them better food and nutrition. A child fully loaded with certain nutrients—especially iron and calcium—is less likely to have ongoing lead absorption and lead settling long-term in their bones. We needed to make sure every child in Flint had a regular pediatrician or "medical home," where they could be treated and their progress reported. This medical home could check children's development and coordinate with all the other programs, and help direct families toward other programs, like home visiting, breastfeeding and parenting support, early literacy, mental health, and transportation access.

The list kept growing. Kirk and I polished it, rewrote it, looked it over and over. We got it down to seven pages, so it was easy to read, the interventions organized by category, then prioritized. We hoped that would make a difference.

And then I got the idea to add a registry to the list. A registry of everyone exposed to the crisis, like the one for the victims of the PBB contamination in 1973, when nine million Michiganders had been exposed to a toxic flame-retardant chemical that causes permanent hormone disruption. Another prime example is the World Trade Center Health Registry, used for victims of 9/11—now the largest registry in the United States to track health effects of a disaster. Like those, the Flint registry had to include people who had moved out of the city, but our focus would be improving outcomes and supporting victims, not just monitoring. It would be a way to make sure that everyone was able to benefit from the advantages and resources that I was going to fight for—and I hoped would come.

Our document wasn't just a list of demands on behalf of Flint kids: it was also a road map to intervention, throwing together everything that we knew worked for kids. And with it in hand, we hoped to enable government and donors to know exactly what was at stake—and what could be expected. We were already almost two years behind for Flint's kids: two years of critical brain development had passed while they were consuming lead regularly. Whenever I thought of that, a sense of urgency took over, and I became single-minded about getting these programs funded and in place.

As the media focus on Flint sharpened, and I had more chances to speak out, I saw an opportunity. As the saying goes, "A crisis is a terrible thing to waste." It wasn't enough for me to give interview after interview, explaining how horrible the state's actions had been, how tragic lead is for children, and how scary the implications were. More could be accomplished with a slightly different perspective from these scare-and-doom tactics. And that was to provide a sense of hope.

And it was with hope—and a commitment to justice—that MSU and Hurley stood beside me and helped me create something new: a model public health program to bring hope and healing to Flint. Coming up with that plan was like building an airplane in midair. But without a moment of hesitation, MSU's Dean Dean and Aron Sousa, along with Melany at Hurley, helped me put it together. MSU made it clear that they would spare no effort to tackle the crisis. Our list of interventions grew and came to include continued advocacy, robust evaluation, and a promise to share our best practices with other struggling communities.

By this time, the immense generosity of people in the United States had been directed at Flint. Truckload after truckload of bottled water was delivered—sent by Girl Scouts in Ohio, UAW locals in Chicago, a ballroom dance club in Ann Arbor. Union plumbers volunteered countless hours installing and replacing filters, and joined the campaign to replace our nation's millions of lead pipes.

The outpouring was tremendous. But soon we'd need more than bottled water. We would need an investment in the tomorrows of our children. Knowing that some of the effects of lead exposure might not be seen for ten years or more, Representative Kildee started talking about setting up a health fund of some kind, separate from the interventions that we were recommending.

In a small meeting, we decided that the best and most transparent place to house this kind of fund would be a community foundation. Kirk and I drafted the framework for a "tomorrow fund" with a twenty-year time frame. We created an advisory committee, established at the Community Foundation of Greater Flint, and I became the founding donor. We called it the Flint Child Health and Development Fund (FlintKids.org), with a goal of raising $100 million.

In January, once Flint had gained "federal emergency" status, we officially submitted our comprehensive science-based list of recommendations to the Emergency Operations Center. That is, we literally put this plan in front of everybody—the city, the state, and eventually the governor. Then we put it in front of the federal government.

This is what Flint kids need, we said.
This is what has to be done.

—

CHAPTER 25

Truth and Reconciliation

I N EARLY FEBRUARY 2016, I WAS SITTING IN THE CLINIC, TRYING to catch up on my backlog of work, when four Michigan state police officers barged in. They had earpieces and dark suits and said the governor was going to pay me a visit.

But before he arrived, a team of guys with wands and all kinds of security gizmos swept and examined every nook and cranny of the clinic. *What the hell?* Most of the staff was shocked, but Arabs are used to this special attention.

Governor Snyder didn't come alone. He was joined by a posse—Lieutenant Governor Brian Calley, who was in Flint a few days a week by then, MDHHS director Nick Lyon, and two indistinguishable top-level aides. Hurley is a diverse place—lots of women, lots of people of color, and a lot of white doctor coats. Now my tiny office was suddenly very crowded with dark suits, worn by five guys who were very tall and very white. Somewhere, down near the floor, were me and my medical student Joe. I felt like Bilbo Baggins when the dwarves keep showing up for the Unexpected Party at the beginning of *The Hobbit*.

The governor started off with an apology. By then, he had been doing a lot of that. Three days after President Obama declared a federal emergency in Flint, the governor had apologized to the people of Flint in his annual State of the State address. He had even used the words *toxic stress* and activated the National Guard to distribute water. A few days later he returned some executive power to Mayor Weaver.

Apologies are easy, however, and in Flint they were starting to come pretty cheap. The minute I started hearing one, I thought, I don't care about your apology. I care about action. I cared about getting enough funding, and the interventions, and how things would work out for the kids of Flint for the next few decades. I would have to work with this governor and many governors to come.

As he apologized, I looked him right in the eye. Governor Snyder had genuine remorse and regret in his voice. He was visibly shaking—I thought he might cry—but he steadied himself and continued. He said he was about to announce the new state budget and wanted to tell me his plans for Flint. The state budget director was on speakerphone. Almost everything I asked for, my entire list of demands—all the wrap-around services for the kids that I had recommended—were, in some shape or form, in the budget.

Nutrition. Health. Education. That meant enhanced meals at school. That meant medical home access for all families and children. Even more support would come from the federal government when it granted Flint a Medicaid waiver. That would mean more school nurses—ten for the city instead of just one. That would mean nurse-family partnerships, home visiting, universal early intervention, school-based health centers, preschools, and expanded pediatric behavioral health access. More than a hundred million dollars of the Michigan budget would be dedicated to Flint kids.

This is the stuff progressives dream about but never get. This is what Alice Hamilton fought to build—"wrap-around services"—when she worked at Hull House. It wasn't long-term money, forever money—a binding budget cannot be forced on future legislatures—but it would begin to help Flint kids now. It was a big deal. I was truly thrilled and surprised. And I thanked Governor Snyder. It was a good start.

Then he had a favor to ask. He didn't exactly put it that way, but that's what it was. He wanted to know if I could attend a press conference the following week when the budget was announced.

I knew what that meant: more trophy time. My earlier head-shaking episode had given me even more credibility. Now he hoped I would stand next to him in public and absolve him of wrongdoing or negligence. If I appeared to forgive him, maybe everybody else would.

Fortunately, I had a legitimate excuse to say no. I was going to be tied up in D.C. that day. I was called to testify about the Flint water crisis to a joint committee of Congress. I was very relieved it worked out that way, even if I was thrilled that he had gone along with most of our demands and suggestions.

Prior to the crisis, Snyder had been popular and had even been considering a presidential run. His fall from grace had been sudden and steep—and completely a result of Flint. His administration had ignored so many red flags. His staff and appointees had ignored the activists' loud and sustained protests. They had ignored the concerns of health professionals and water treatment experts. They had chosen to look the other way, and when faced with the results of their own apathy and negligence, they had tried to cover up.

If I had to locate an exact cause of the crisis, above all others, it would be the ideology of extreme austerity and "all government is bad government." The state of Michigan didn't need less government; it needed more and better government, responsible and effective government. *And people who are not being poisoned by the tap water.*

For decades, the city and state infrastructure had been neglected in order to save money. State environmental and health agencies had been defunded, and great public servants had become disillusioned and retired, leaving these agencies a shadow of what they were supposed to be. All the budget cuts and so-called fiscal "responsibility" had resulted in a winner-take-all culture, a disdain for regulations and career regulators, a rubber-stamping of bad ideas, a gross underfunding of environmental enforcement, limited understanding of and expertise in public health, and a disregard for the poor. Snyder's people tried to be good at PR, but in the end they even failed at that.

The implementation of the EM law was the deciding factor— evidence of contempt for Flint and its poor residents, as if they were subcitizens who didn't deserve a fair say or vote. Years before the water switch, members of the state's black caucus in Congress had asked the Justice Department to review the constitutionality of the EM law. Though they were ultimately unsuccessful, residents of cities taken over by emergency managers fiercely argued that the EM law violated the Voting Rights Act, since it subjected more than half of Michigan's African-American voters to emergency management in their schools or cities.

"Everybody knows that this [water crisis] would not have happened in predominantly white Michigan cities like West Bloomfield, or Grosse Pointe, or Ann Arbor," Michael Moore wrote in *Time*. "Everybody knows that if there had been two years of taxpayer complaints, and then a year of warnings from scientists and doctors, this would have been fixed in those towns. . . .

"This is a racial crime. If it were happening in another country, we'd call it an ethnic cleansing."

ABOUT A MONTH AFTER my press conference, Governor Snyder named five men to a Flint water task force. When I first heard about it, I was skeptical—not just because they couldn't bother to find a single woman but because naming a blue-ribbon committee is a time-honored way to "solve" a political nuisance. But I was reassured to hear that Dr. Reynolds was on the task force, and that it would be cochaired by former state representative Chris Kolb, an environmentalist and also a student of Bunyan Bryant. Kolb had been prematurely expelled from public service because of Michigan's stupid term-limit law. I started calling it the Five Guys Committee, and not long afterward the task force informally adopted that name for itself.

It was charged with "reviewing actions regarding water use and testing in Flint" and asked to "offer recommendations for future guidelines to protect the health and safety of all state residents."

The Five Guys took their mission seriously, probably more seriously than the governor intended, and in March 2016 released a kickass 112-page report. It fixed blame for the failure of government primarily on MDEQ, but it didn't spare the governor, the emergency managers, or the MDHHS. The report stated very clearly that the demographics of a community—race—had played a role in creating this environmental crisis. And it stated that race had kept the crisis going long after it should have been stopped. The Five Guys explicitly connected the Flint water crisis to the emergency manager law. And in this way, the report was historic. Never had a government report explicitly stated the role of race in an environmental crisis. And never before had the consequences of the loss of democracy been so keenly demonstrated. Flint may be the most egregious modern-day example of environmental injustice.

Later, in February 2017, the Michigan Department of Civil Rights made similar findings in its own investigation. Another father of environmental justice, Robert Bullard, once said some communities have "the wrong complexion for protection." That is how I see it too.

Flint falls right into the American narrative of cheapening black life. White America may not have seen the common thread between Flint history and these tragedies, but black America saw it immediately. That the blood of African-American children was unnecessarily and callously laced with lead speaks in the same rhythm as Black Lives Matter, a movement also born from the blood of innocent African Americans. Trayvon Martin. Eric Garner. Michael Brown. Sandra Bland. Freddie Gray (also a victim of lead poisoning). Alton Sterling. Tamir Rice.

At a January 2016 rally, Jesse Jackson called Flint a crime scene: "I guarantee you if it was any of my homeboys, my friends, they'd be in handcuffs right now." Think about it. How many African Americans, and others, are in jail for the most minor offenses, simple assaults, minor burglaries, bail violations, drug crimes? One in three

African-American men will be incarcerated at some point in their lives. Too many. It's racially driven mass incarceration. Yet I won't be surprised if relatively little jail time is served by anyone for the Flint mess, even though it could measurably alter eight to ten thousand kids' lives.

I was prepared for years of battle and denials. I never thought the governor or his administration would be capable of comprehending and accepting the charge of willful negligence—or would ever do anything about it. But I was wrong: when faced with data that could not be ignored and the full weight of the crisis—the protesters and the media onslaught—he got his act together. Though his words were stilted, too slow, and never robust enough, his deeds, and the deeds of his administration, deserve some respect.

I'm wise enough to know that Brad Wurfel and others got their marching orders from higher up, possibly all the way up. I also know that most people, except for megalomaniacs like Saddam Hussein, don't wake up in the morning saying, "I want to poison thousands of people."

Admitting your mistakes, and then doing what you can to rectify them, takes integrity and strength. And in the end, I felt the governor cared—and was truly sorry. This gave me hope, because a man who cares and feels sorry might do more for Flint kids, to rectify his mistakes, than a politician with less at stake. Even if he didn't care and wasn't sorry, he knew that the balance of his political career would be judged by his response to this crisis.

That's why, rather than seeking retribution or political advantage, I decided to remain loyal to only one group: Flint kids. I will do whatever it takes to help them and to keep on helping them, even if it means working with people whose ideology and actions got us into this mess.

At the end of my meeting with the governor, he asked for my blessing, and I gave it to him. Four months later, in June 2016, after a lot of dogged advocacy, the Michigan state legislature, following the lead of senate minority leader Jim Ananich, passed the budget.

The three state budget supplements, passed at three different times, totaled about $200 million in aid for Flint.

It would take another six months of coalition building, roller-coaster ups and downs, and trips to D.C.—more than a year after the crisis was exposed—before a Flint aid package finally passed in the U.S. Congress, with much thanks to the stubborn efforts of Congressman Dan Kildee and Senators Debbie Stabenow and Gary Peters, as well as Congressman Fred Upton, a Republican from western Michigan.

The advocacy life is like that. One day, a victory. The next day, a crushing disappointment.

And the next, another victory.

I DIDN'T MEET MIGUEL DEL TORAL until he came to town in March 2016. We had emailed and talked on the phone, but we had never met until just after President Obama declared the federal emergency, paving the way for more EPA resources and intervention. So our meeting came at a happy moment, in the glow of one of those victories.

We were all supposed to have dinner—Marc Edwards, LeeAnne Walters, Miguel, and me. But Miguel was stuck in his hotel room with severe back pain. We all had gone through dark nights of the soul as advocates for Flint. But the efforts to silence and ruin Miguel for his whistle-blowing—and his being dismissed, having to leak his own memo, and being sent to the EPA ethics office—had taken a toll on him physically. He could barely stand, let alone walk. So we all went up to his hotel room, just to say hello. I really wanted to meet him, to thank him, to hug him.

When we were introduced, Miguel leaned over and kissed me on the forehead, the endearing way you'd greet a child. I still remember the way that felt and how sweet it was.

You can't tell from the photograph, but we are all holding him up—and helping him stand. When we helped him sit again, I noticed that his feet were bare and his toenails were painted. Through-

LEFT TO RIGHT: LEEANNE, MIGUEL, ME, AND MARC

out the crisis, his wife had complained that Miguel was so consumed by the Flint water, so infatuated with lead pipes, that he had stopped listening to her and couldn't remember anything she said. So one night, while he was on the phone talking about the water, she painted his toenails to get him to pay attention to her—and it worked. The polish was flaking off but still there.

A MONTH LATER ELIN and I had lunch, meeting at our favorite diner. It was April, and a lot was still happening in Flint. The final Five Guys report had come out the month before, spelling out in harsh terms the negligence and willful disregard for health, particularly by MDEQ. Brad Wurfel had lost his job in December, right after Christmas, the same day his boss, Dan Wyant, resigned.

In the past six months, more than fifty lawsuits had been filed, and the first criminal charges were announced—against two state officials, Stephen Busch and Mike Prysby in the drinking water division of MDEQ. There were six charges altogether, including misleading the EPA, manipulating water sampling, and tampering with reports. MDEQ received the most indictments and charges. Besides Busch and Prysby, the top drinking-water regulator, Liane Shekter-Smith, was fired.

Three members of the MDHHS epidemiology and lead staff— Corinne Miller, Bob Scott, and Nancy Peeler—were charged with actively covering up the spike in blood-lead-level data. I never learned why nurse Karen Lishinski failed to follow up with me. She was never accused of wrongdoing, but her superiors at the state lead surveillance program were.

Down the road, MDHHS director Nick Lyon and Eden Wells were charged, related to the Legionnaires' disease outbreak. Wells was charged with obstruction of justice, Lyon with misconduct. Both were charged with involuntary manslaughter, along with three others—Busch, Howard Croft, and Shekter-Smith—for their failure to act in the crisis.

The criminal complaint against Lyon included a reference to his email to subordinates, asking them to make "a strong statement with a demonstration of proof that the blood-lead levels seen are not out of the ordinary."

Susan Hedman, the director of EPA Region 5, where Miguel worked, resigned in January 2016. A scathing report by the agency's inspector general said that the EPA should have acted seven months sooner. This prompted Representative Kildee to lead a charge to pass a bipartisan bill that requires the EPA to notify the public any time lead is found in the water supply.

Flint officials weren't spared. City manager Natasha Henderson was fired in February; Howard Croft lost his job and was charged with two additional felonies—false pretenses and conspiracy to commit false pretenses in connection with the water crisis. Another Flint city official, Mike Glasgow, was accused of tampering with a lead report. Later, Flint emergency manager Darnell Earley was charged with involuntary manslaughter.

Dayne Walling, who lost his bid for reelection just six weeks after my press conference, was the first political casualty of the crisis. I have often suspected that if he'd been on the right side of the fight—and had skipped his chance to meet the pope—things would have been different. Today, he regularly expresses regret about his role.

Governor Snyder survived his own crash in the polls and was not forced to resign. But his reputation is tarnished, and he appears to have abandoned thoughts of seeking higher office.

Many people asked me, "Shouldn't the governor go to jail?" Truth and reconciliation and restorative justice are part of healing. And investigations are important. But the law is not my job. My focus is on the health and well-being of Flint kids.

Accountability is another matter. In D.C., it is what Marc Edwards hoped for but never received. The D.C. water crisis took a decade to be resolved, and unsatisfactorily: negligent and deceitful government employees and water utility executives never paid a price, were never charged with a crime, never even lost their jobs. Most of the children in D.C. who were affected—potentially 42,000 of them—were never compensated or even made aware of what had happened. When Marc came to Flint, he despaired that his work wouldn't make a difference, even though he was committed to doing all he could. It was this sense of resignation, and even bitterness, that I saw in his eyes and face the first time we met in September. And when he looked across the table and said, "I trust you," the leap he was taking was obvious. This was a man who had been undermined, lied to, and dismissed by public health "experts" for years, but who never gave up on his mission.

After our initial meeting at the farmers' market, hundreds of emails passed between Marc and me, both curses and celebrations. He became a fixture in my life, like a long-lost brother. But it would be another two months before we met again face-to-face. We were both in D.C. for a meeting of the EPA drinking water committee at which the existing Lead and Copper Rule was up for discussion. It was our chance to advocate for strengthening the rule. We wanted the EPA to finally force utilities to methodologically optimize corrosion control and replace the lead pipes.

On our way to the hearing, Marc was stopped by a woman who recognized him. She had graying hair and looked worn and tired. She said she was an activist in D.C. and had a son who was a baby during the lead exposure. A teenager now, the boy's learning and behavioral problems were significant. "There is not a day that goes by that I don't wonder if it was the D.C. water that impacted my son," she said, then thanked Marc for his work.

As a mom and a pediatrician, my heart ached for her. I wondered when we were going to learn from our past mistakes, when our policies would finally catch up with science. Science tells us that there is no safe level of lead. All the federal lead exposure standards are out of date. Rulemaking for something so crucial should not take almost a decade. Every day children are trapped and held hostage by bureaucratic governmental processes.

We know the right policies and practices to eliminate lead exposure for all children. And we even know the economic benefit. But we continue to kick the can down the road, putting the future of our most vulnerable kids at risk.

As Marc and I walked on, I kept thinking about that activist mom. And I knew that years from now Flint moms would surely wonder the same thing: *How much lead was my child exposed to, and what did it do to her?*

At the hearing, the EPA allowed us to make a case for a stronger Lead and Copper Rule in two-minute allotments. A bunch of us were there—Representative Kildee, Senator Ananich, LeeAnne Walters, Curt Guyette, and others. After the testimony, Marc and I headed to Capitol Hill to continue our advocacy. We first went to Kildee's office for a meeting, where we ran into the legendary water warrior Erin Brockovich, and then we went on to see Senator Gary Peters before going our separate ways. Marc had a class to teach remotely.

I had a meeting with Senator Stabenow, who had kept her

promise and, in record time, paved the way to make ready-to-feed formula available for all Flint babies, like Nakala.

Just before we parted, Marc said, "Can I speak to you?" and pulled me into a small room in Senator Peters's office. During the day, I had noticed he seemed a little strung out and preoccupied. Aside from his two-minute testimony to the EPA, he had been quiet. Was it hard for him to be back in D.C., I wondered, where he undoubtedly had so many bad memories?

"You are amazing at this," he said as soon as we were alone. "Do you know that?"

I was embarrassed and wanted to stop him, deny what he was saying, be humble, dismiss it, change the subject, and remind him that this was part of my job. This was what pediatricians were supposed to do, what we take an oath to do.

But I'll confess, it also felt good to bask in the sunshine of this compliment, coming from a man who had been in the trenches, been an activist for kids, fought harder and longer than I had—and put more on the line. What he was saying, really, was that I could do things he couldn't: spread hope, bring people together, control a room. His heart and his science were in the right place, but he wasn't an organizer or an optimist.

It wasn't really what I had achieved that impressed him. It was what we had all achieved together. We were, as he said, "a dream team."

It wasn't lost on either of us that we were lucky to be able to speak out—that we had the support and academic freedom of the public land grant universities where we taught and did research. Both Michigan State and Virginia Tech had our backs and urged us on.

Marc continued, probably sensing how much it meant to me. I connected with audiences, he said. I connected emotionally—I made people feel engaged and angry and ready to help, ready to fight. "You are a force of nature," he said.

Not long afterward he gave me a necklace made from a slice of lead pipe that came from a D.C. home where a child had high lead levels—and the city kept denying that any plumbing in the house contained lead until the pipe was dug up. When he gave it to me, plated in gold, he said he hoped it would be a constant reminder of my commitment to all kids.

I keep it with me always, wherever I go.

ELIN AND I ORDERED our lunch and chatted in that casual old-friends way about parents, kids, schools. Sitting across from her, I noticed she seemed lighter. Her manner was easier. She was more like her younger self, her high school and college self. Her years in D.C.—the dismay, the disturbing realization that "environmental protection" can mean all kinds of things, including butt-covering, pointless exercises, and pro-industry agendas—had left her burdened and puzzled. She had carried around a sense of disappointment in life, a kind of existential dissatisfaction. But now, that heavy air around her had cleared.

Eight months had passed since the noisy barbecue at my house where she cornered me to tell me about the water in Flint. It seemed like a lifetime ago. It almost felt like we were both new people now, older and wiser, but also younger and more hopeful. I had spent the better part of these months in meetings, giving interviews and speeches. I had seen Marc many more times over the winter and spring, at hearings, Flint water meetings, and various functions where we shared our accounts of the Flint story. *Time* magazine cited us in its annual list of the world's most influential people, and, most memorably for Mama Evelyn, I received a Detroit Pistons jersey with my name on it during a halftime show.

All the recognition was largely awkward. I often feel like a broken record, always repeating that I was doing my job, adhering to an oath I had taken when I became a doctor. And that my activism on behalf of Flint and the kids of Flint is a privilege, full of its own rewards that have nothing to do with awards. Although I wouldn't wish the state's attack machine on anyone, my woes are nothing compared to the daily deprivations inflicted on the people of Flint, who, at this time, still do not have drinkable tap water. They are the heroes of this crisis.

What I liked most was the chance to explain, to share the latest developments in Flint, to tell the stories—and to track the lessons. I hope that somehow what happened—the resistance, the science advocacy, and the coalition building—inspires others to fight the injustices around them.

"I told you," Elliott liked to say. "I told you that you could do it—that you'd find a way to fight for Flint kids. And you did."

"You never said that," I liked to respond, as stubbornly as possible. And it always made him laugh.

In early March, he'd had another operation on his shoulder—to repair the mistakes made in the botched first surgery—and he was recovering quickly this time, and making plans to build a tree house in our backyard.

Nina and Layla still complained that I was gone too much—and traveling too much—but they were happily obsessed with Simba, an emotionally needy rescue cat that followed them all over the house, never leaving them alone.

Marc stayed with us during his frequent pilgrimages to Michigan, where he served on various committees and continued water sampling. By then, he wasn't just my close friend; he was a good friend to my entire clan—my brother, the girls, Simba—and was carefully monitoring Elliott's progress on his treehouse plans.

And even though the road to clean tap water in Flint would take years, and the people of Flint still had a lot of suffering ahead, Marc always says that what happened here—in contrast to D.C.—was a "miracle," something he would never have believed possible.

Doctors aren't really supposed to be mystical, but sometimes the whole thing had a "meant to be" feeling, starting with the wild coincidence that Elin had been right there at the EPA throughout the D.C. crisis.

Water treatment engineers aren't supposed to be mystical either. "Before Flint, my career was a total mystery to me," Elin confessed at lunch. "None of it made sense. So many times I have looked back at my choices and career moves and wondered, *What was I thinking?* But now I see that each move was a piece in the puzzle, so that I could be standing right there in your kitchen at exactly the right time. It all makes sense to me now. There was a point to all those twists and turns."

This was a daunting revelation coming from a woman so confident and strong. But the more I thought about it, the more I realized that Elin's description really described my own life. Like lots of young people, I was drawn to advocacy, but the chance to really do something meaningful had come by accident. I was the right person in the right job, with the right set of skills, with the right team, who saw a chance to make a difference. For me to push through and succeed, many random aspects of my life came into play and helped me, from the resilience of my long-dead ancestors and the courage of my progressive family to my many mentors who fostered my environmental activism, which started in high school and grew in college, and my background in hard-core science.

The Flint water crisis never should have happened. It was entirely preventable. And when the water supply was switched and not properly treated, it should have ended when the residents raised their voices and their jugs of brown water. I certainly wasn't the most important piece of the Flint puzzle. So many others—LeeAnne, Miguel, Elin, and Marc, the dozens of tireless Flint activists and local journalists—were instrumental and critical.

I was just the last piece. The state wouldn't stop lying until somebody came along to prove that real harm was being done to kids. Then the house of cards fell.

I FINALLY HAD A chance to talk to Bunyan Bryant, the environmental justice pioneer, when I received the Michigan Environmental Council's Distinguished Service Award, named after environmentalist and former Michigan governor William Milliken. The last time I'd seen Bunyan was more than a decade earlier at my brother's

**PRESENTING AN AWARD TO BUNYAN BRYANT, FLINT
ENVIRONMENTAL JUSTICE SUMMIT, MARCH 2017**

wedding to Annette, an environmental lawyer who's even closer to
Bunyan.

Bunyan's family is still in Flint, and his lifelong fight for justice
buzzed through my mind as I talked a mile a minute about the Flint
water crisis with him. He is suffering from Parkinson's now. His
wife tells me he has good days and bad. I had the great fortune of
seeing him on a good day. Always the professor, he had some keen
observations to share. I knew he was thinking about all the places
plagued by ongoing environmental injustice—including the many
communities with lead contamination much worse than Flint's. He
couldn't quite respond a mile a minute, but I knew his sharp mind
was taking everything in. Bunyan's lifetime achievements and contributions in environmental justice gave me, and all of us, a way to

see the Flint crisis through his eyes, and a framework and language to understand it—truly an enduring legacy.

Mostly, I wanted to thank him. Throughout the Flint crisis, I stood on the shoulders of giants—incredible giants I intimately worked with, like Marc Edwards, and giants who walked a long time ago, like Alice Hamilton and John Snow, who paved the way. Bunyan Bryant is definitely one of them.

—

Prescription for Hope

N AKALA CAME TO THE CLINIC FOR HER EIGHTEEN-MONTH checkup in October 2016. Reeva spotted me in the hallway while her baby sister was being weighed. "Dr. Mona! We saw you on TV!" I waved her over, lifted her up, and squeezed her tight. Holding her hand, we walked over to the bookshelf and looked at newly stocked *Reach Out and Read* books. We grabbed a couple, and a Richard Scarry classic for Nakala.

Back in the exam room, I smiled at Grace, their young mom. She seemed more grown up, more confident somehow. She and I had more of a relationship than I have with most of my patients. I met her when Nakala was just four months old, the week I learned about the Flint water, and somehow those memories are seared together in my mind. When I thought of the babies of Flint, I always thought of sweet Nakala, who was now a chubby, happy, walking-and-talking toddler.

I had done a lot of apologizing to Grace a year earlier. "I told you the water was fine for Nakala's formula, and it wasn't," I had said. "I'm so sorry. I had no idea." Moms are always right. Every parent should listen to that primal instinct to protect their kids.

"I knew it wasn't fine," Grace replied, much to my relief. "My auntie told me. She's Nakala's godmother, and her water was brown and smelled. My water seemed okay, but my auntie told me not to use it. I said, 'I can't afford bottled water,' so she bought it for me."

Allison and I had seen Nakala for her checkups when she was six, nine, twelve, and fifteen months old. Over the year, Grace mixed her formula with bottled water that she got from her auntie, then from her church, the food bank, and the National Guard. She cooked Reeva's meals with it too.

The medical assistant came to take Reeva for her blood draw. She was getting her lead level checked. Her first lead level, taken the year before, was high—26 μg/dl—and now she was getting it checked again. After that, she'd have her hearing and vision checked, and get a ten-dollar prescription for fruit and veggies at the Flint Farmers' Market, which becomes twenty dollars with the Flint expansion of Double Up Food Bucks.

As soon as they left, Grace let down her guard and gave me an all-too-familiar look of fear, anxiety, and guilt. "Is she going to be okay?" she asked. "Both of them. Are they going to be okay?"

Reeva was three years old that autumn. She had been drinking contaminated water most of her life, during the most critical period of brain development. And even though Nakala was spared the water in her formula, Grace had drunk it during her pregnancy.

A year before, their water at home had been checked and found to be 150 ppb, something Grace would never have known on her own, because lead in water is colorless, odorless, and tasteless. At their auntie's house, it was 3,000 ppb, and a neighbor down the street had a level of 13,000 ppb. Putting this into perspective, 15 ppb is the federal action level (and that's set too high)—and 5,000 ppb is the hazardous waste level. Drinking Flint water after the water switch was like drinking through a lead straw. You never knew when a piece of severely corroding lead scale would break off and fall into the drinking water and into the bodies of our children. When I found out that a group home for abused and neglected kids near the hospital had a water-lead level over 5,000 ppb, I was the maddest and saddest I'd ever been in my life. These kids literally had every adversity possible. It was like the world was conspiring to keep them down.

Grace worried a lot. And there were days when she looked terrorized and down. I couldn't blame her. Every day in Flint, our families had been waking up to a nightmare of water contamination, betrayal by every governmental agency charged with protecting them, and years of neglect and poverty. Many of the families who want to call it quits and leave Flint can't sell their homes, because the home values have tanked. Those who remain feel angry at everyone almost every day. It is community-wide PTSD. The mental health problems are now just as serious as the physical ones.

My FAMILY CAME TO the United States basically as refugees fleeing oppression, in search of a peaceful and prosperous place for my brother and me to grow up. The American Dream worked for us.

But sitting with Nakala in my lap, I realized that America has changed a lot since I was a little girl. Yes, people are still running to America, or at least trying to. It remains the epitome of prosperity for the entire world, the richest country that ever was. But there really are two Americas, aren't there? The America I was lucky to grow up in, and the other America—the one I see in my clinic every day.

In that other America, I have seen things I wish I'd never seen. The things you run from, not toward. Things that would never be part of any dream. And for too many people, this nightmare is taking place right outside their front door.

Sometimes I joke that I was born an activist. But it's not really a joke. I was born into a family that was on the move toward something better, and I was born into a life knowing there is injustice in the world—and the importance of fighting it. And that's exactly what my babies in Flint are born into—sweet little ones like Nakala. For them, life is a struggle from the very beginning. That can make a baby a fighter. Because for Nakala and Reeva, every little thing—sometimes things as simple as a meal or clean water or a bath—can require a fight.

Where is the American Dream in this scenario? It's not there. It's not even talked about. It is becoming so out of reach. Income and wealth inequality make mobility tough. Stagnation is now the norm. At the end of their lives, most children wind up where they started. Not just Flint's kids but children in Detroit, and Los Angeles, and Chicago, and Baltimore, and all over rural America. Black, brown, and white. Too many kids are growing up in situations stacked with insurmountable toxic stresses and every barrier imaginable. Too many kids are growing up in a nation that does not value their future—or even try to offer them a better one.

That's not how it's supposed to be. The Dream shouldn't have to come by way of a miracle. It should come fairly to all and be big enough for everybody to achieve. The environment shouldn't be stacked against anybody, especially our kids. We owe it to them and to one another.

As a doctor, what can I prescribe for our sons and daughters of the other America?

This is what I thought about that day with Grace, when she asked me about her girls, and if they were going to be okay. Nakala was still snuggling on my lap.

"What can I do?" Grace asked.

"Keep doing everything you are already doing. Love them, talk to them, sing to them, read to them, give them great food," I replied. I encouraged her to fill the "nutrition prescriptions" downstairs at the farmers' market for fresh produce, and urged her to sign up for WIC. There is a special Medicaid expansion and universal early-intervention development support offered to Flint kids. "You have also been getting the home delivery of kids' books every month, right?" Mental health support is nothing to be ashamed of, I said, "so here is the number to the trauma crisis line if you need it. Don't forget about the mindfulness workshops—they have them in the schools now too. Parenting support classes are being held on Tuesdays, and they even do home visits." I took a breath and said, "And make sure you come visit me often."

Grace applied for a preschool aide job at Flint's new childcare center, one that pays well, more than $17.50 an hour, with healthcare benefits and a 401(k). It was a new position made available with state money. She was sticking with Flint. This "intervention" of a good job, a living wage job, is one of the best I can think of. And these recovery jobs, the return of the Grand Bargain, will be critical for Flint and places like Flint everywhere.

"Nakala and Reeva are growing so fast right now," I went on. "Their brains are growing so much. And no matter what badness they may have been exposed to, we can overcome it with all this goodness. We are trying to tip the scale. And we will."

The most important medication that I can prescribe is hope.

Nakala moved around in my lap, a little restless. Her eyes were wide open, and she followed my voice and my every move. Her fingers were tightly wrapped around mine. I lifted her up and told her she was going to be fine. She was mixed up in an accident—a lot of kids were—but it wasn't her fault. And it's my job to make sure she and her sister—and all the kids in Flint—are okay. Actually, it's my job to make sure they have every chance in the world to be better than okay—to be great.

Haji and the Birds

ONE NIGHT I'M WORKING LATE, AND I HEAR BEBE HELPING Elliott get the girls ready for bed.

Nina says she isn't sleepy and never wants to go to bed. Layla pretends not to hear. She is dressing Simba. Poor cat. There is a long silence, followed by movement in the hallway. Bebe, in her quietly persistent way, inches them toward sleep by offering to tell them a story.

Nina doesn't really need anybody to read to her anymore. She is a bookworm in her own right. Elliott and I still love the bedtime reading ritual, though, because there is nothing like bedtime cuddling while bonding over a story. And research on early development has shown how critical it is to little kids. Reading to kids is brain food. But this year, as well as last, Elliott still has the most chances to put the girls to bed. It's something I miss.

As soon as the girls settle in, I hear Bebe's gentle voice begin. She isn't reading a book. She's telling the story of *Haji wa al asafer.* "Haji and the Birds." It's a family fable that has been passed down, a story my mom heard from her father, Haji, when she was little, and later used to tell Mark and me. It is about a young man named Haji, but we all know my grandfather was that young man—which makes it more vivid and intense.

"Layla, you don't remember your great-grandfather," Bebe is saying in Arabic. "But Nina does, don't you?"

"I've seen pictures of him, Bebe," Layla calls out in her husky voice. "I know him!"

"He was a special person," Bebe goes on. "When I think of him, he gives me strength. He was honest and kind. He was generous."

"Oh yeah!" Layla calls out. "I remember the story Mama told us about how he helped his workers and then they helped him!"

"That's right," Bebe says. "Haji loved everybody. But especially he loved kids. He had five kids, so there was a lot of love in my house when I was a girl your age.

"At night in the summer, instead of sleeping in our bedrooms, everyone slept on the roof, the *tarma,* where it was cool. That's what people in Baghdad used to do and still do."

"I wouldn't like that," Layla says.

"Maybe you would," Bebe answers in Arabic, "if you tried it. And before we all fell asleep, Haji used to cuddle with us and tell us this story."

"Haji Jassem and the birds!" says Layla.

"That's it!" says Bebe. "You have heard it before. But you need to hear it many times, because someday you'll be telling the story to your own children, and your grandchildren. They'll all come to know Haji that way, and the things he cared about and thought were important."

After a short pause, she starts. "Once upon a time in Baghdad, there was a young man named Haji. He loved nature. He loved flowers. At his house, he had a garden of citrus trees and date palms, pomegranate trees and flowers.

"Most of all, he loved birds. He loved their freedom. He loved the sounds of their singing. Black birds, white birds, yellow birds. Big birds and very little birds. Every morning he opened the window to hear their different songs. He loved the cooing of pigeons. He loved the chirping of little sparrows. Each song was different and separate, but together the birds made a beautiful symphony.

"Every morning, he walked outside with a bowl of grains and

rice and seeds. He made a feast for the birds. They would fly down, and Haji liked to watch them eat his feast, then drink from the flowers. He became friends with them all; some birds even ate from the palm of his hand. The garden was such a simple and peaceful place.

"One day, after feeding the birds, Haji climbed a ladder to pick dates from a palm tree, and reaching for a branch slightly too high, he fell off and broke his leg. He called out for help, called and called, but nobody came.

"A small bird flew down and tugged at the hem of his white *dishdasha*. The bird told Haji that he would take him to the doctor. But Haji laughed at the small bird, wondering how such a tiny bird could carry him. Soon another bird came and took the edge of his sleeve. Another bird came, and another, until hundreds of birds surrounded him. They each held a small piece of his *dishdasha*, and even his hair and his toes, and together the birds were able to lift him and fly him through the air.

"They flew over the palm trees, the majestic meandering Tigris, the statues of the ancient poets and caliphs, and the impossibly blue minarets and cramped alleys of old Baghdad. It was a magic carpet of birds. And they flew him all the way to the Baghdad hospital, where a very kind doctor took care of his leg. In a few weeks, after his care, Haji was much better and went out into his garden to see the birds again."

I LISTEN TO MY MOM until the end. And I wonder, as I always do, if Haji thanked the birds. And I like how each time the story is told, it changes a little bit.

When Haji told the story, he liked to draw out details so it took a really, really long time. He described every single bird and every part of Haji's body being lifted up.

When Bebe tells the story, she describes the flowers and trees in the garden, and the picturesque neighborhood in Baghdad where Haji lived, and where she was raised.

When my brother tells the story to his boys, he elaborates on Haji's skepticism that one tiny bird can help him, until he sees how all the birds working together are able to lift him up.

When I tell it, and get to the ending, I have a little more to say about the hospital and the kind doctor who cures Haji's leg. I always say she's a lady doctor. And I remind the girls how Haji treated every person and every being and his surroundings with respect. And he was treated with the same respect in return. Haji took care of people, and they took care of him. It is a simple way to live, the right way to live. It brought Haji much success in life—to give and receive so much comfort and security and happiness.

Haji lived long enough to see me become a doctor—and to come to Mark's wedding and mine (even to the church)—and to meet our eldest kids. This means more to us than we can say. Even though he and my grandmother Mama Latifa were in faraway Baghdad throughout our childhoods, they made their presence known, and their love felt. As we grew up, Haji wrote separate letters to us in Arabic and English, in tiny handwriting that only my mom could read. He knew what books we were reading, especially if they were classics, and what our favorite subjects and teachers were. He knew about our environmental activism in high school, and he championed our commitment to social justice, reminding us of his idealist cousin Nuri Rufail, who risked his freedom and his life to fight fascism.

Haji never wanted to leave Baghdad, to give up his businesses and rich social life—his many friends from every walk of life, every religion, ethnicity, and sect. He never wanted to leave his magical garden or all the birds he fed every day. But eventually he came to Michigan to escape the bombing and danger. And he wanted to be closer to us and his great-grandkids. When they went to bed, he told them how he fed the birds in Baghdad when he was young.

And now, almost ten years after his death, I can still feel Haji and feel his magic lifting me up.

Acknowledgments

I HOPED AND DREAMED THAT THESE PAGES WOULD SOMEHOW weave the stories of my immigrant family with the events of the Flint water crisis to make better sense of both. I had so many seemingly disconnected subjects to share—from drinking-water safety to a genocide in Iraq—but in my mind they were intertwined and connected, all pieces of the same story, the same lens, that explains how I see the world.

This book started out as a three-hundred-page "draft" that I began soon after the crisis exploded. Thanks to the experience, expertise, and wisdom of Martha Sherrill, it began taking shape and coming together. Thank you for guiding, translating, and massaging my rough words and ideas—and for not telling me the "draft" I handed you was a bit of a disaster. I'm still flummoxed and grateful that you embraced this genre-defying puzzle and made it part of your heart too.

Like so much in this serendipitous story, it was fate that I ran into Chris Jackson in D.C., where he asked, "Have you thought about writing a book?" Thank you, Chris, for believing that what I had to say belonged in the same inspiring songbook as the other important social justice stories you have helped conceive. Your creativity, insightful perspective, and hands-on involvement lifted every theme in this book. I'm grateful to your entire team at Penguin Random House and the One World imprint—Nicole Counts, London King, Matthew Martin, Loren Noveck, Sharon Propson,

Jessica Bonet, Greg Mollica, and Random House president Gina Centrello. Thank you to Penguin Random House for their donation of children's books to the Flint Kids Read program.

Always energetic and encouraging, my agent, Jennifer Rudolph Walsh at WME, brought leadership and savviness to every step of this endeavor. Jennifer, your positive energy is contagious. Much thanks to Eric Lupfer, who beautifully packaged my book proposal early on, and a shout out to Nell Scovell for giving me an early lesson in the strange world of publishing.

Thank you, early readers: Sinan Antoon, Elin Betanzo, Marc Edwards, Jenny LaChance, Daniel Okrent, and David Rosner. There were other experts I kept bugging— Bruce Lanphear, Rick Sadler, and Nigel Paneth—and want to acknowledge, along with photo and reference researchers Tim Thayer and Katherine Negele, and perfectionist wordsmith John Kenney.

There aren't enough pages to thank all the people who have been important in my life, because there are simply too many. But I would be remiss not to mention the teachers, professors, role models, mentors, and colleagues who have influenced and inspired me. Some of you are discussed in these pages, but many more are not. Together, you made me the scientist, activist, storyteller, pediatrician, and person I am. Thank you to my past and current learners— my residents and students—who have taught me so much over the years, far more than I've taught them.

And behind the scenes in a teaching hospital and university there is a network of support—the administration, nurses, building staff, and so many others: your labor makes teaching, learning, and healing possible. Thank you to the leadership at Michigan State University and Hurley Medical Center—especially Aron Sousa, Melany Gavulic, and James Buterakos—who continue to have my back.

And most important, thank you to my patients and their families, who trust me to care for them. It is an honor to tell some of your stories in these pages. I am fortunate to be the one looking in your little ears, listening to your hearts, and hearing your concerns.

Being your doctor during the crisis, before, and after will never just be a job: it is the privilege of my life.

Now, my family . . . When I laughed off the idea of writing a book—saying "It's not about me" and "Don't I have enough going on already?" and "*How on earth can I find another minute in the day?*"—it was you, my brother and oldest friend, Mark (who also doesn't have an extra minute to spare), who shared this journey with me. You will always be Muaked, and never more certain, more confident.

To my parents, Talia and M. David Hanna—who still take care of me, now even more so—thank you for understanding the importance and value of resurrecting, remembering, and retelling all your family stories. They are woven into me, they lift me up, and they keep me going. I hope this book serves to honor your idealism, your words and deeds, and your lifelong fight for justice.

Finally, to my husband, Elliott—always patient and encouraging: I am grateful for all your creative contributions and attention to detail, especially with the photos, even if I sometimes stubbornly don't listen and refuse to tell you so. And most important, thank you for being my partner as we scramble to raise our girls, Nina and Layla.

To them I'd like to say that I know in the last two years I missed a lot of tuck-ins, drop-offs, pickups, soccer games, Girl Scout meetings, and so much more. Even when I was home, I always seemed to be working or writing the "Dr. Mona Book." Thank you for graciously sharing your mama. Haji's lessons are for you.

Further Reading

===

al-Khalil, Samir [Kanan Makiya]. *Republic of Fear: The Inside Story of Saddam's Iraq.* New York: Pantheon Books, 1990.

Batatu, Hanna. *The Old Social Classes and the Revolutionary Movements of Iraq.* Princeton, N.J.: Princeton University Press, 1983.

Boyle, Kevin. *Arc of Justice: A Saga of Race, Civil Rights, and Murder in the Jazz Age.* New York: Henry Holt, 2004.

Bridge magazine staff. *Poison on Tap: How Government Failed Flint, and the Heroes Who Fought Back.* Traverse City, Mich.: Mission Point Press, 2016.

Burke Harris, Nadine. *The Deepest Well: Healing the Long-Term Effects of Childhood Adversity.* Boston: Houghton Mifflin, 2018.

Carson, Rachel. *Silent Spring.* Boston: Houghton Mifflin, 1962.

Colborn, Theo, Dianne Dumanoski, and John Peterson Myers. *Our Stolen Future: Are We Threatening Our Fertility, Intelligence, and Survival?—A Scientific Detective Story.* New York: Plume, 1997.

Denworth, Lydia. *Toxic Truth: A Scientist, a Doctor, and the Battle over Lead.* Boston: Beacon Press, 2009.

Edwards, Marc. Flint Water Study. flintwaterstudy.org.

Fine, Sidney. *Sit-Down: The General Motors Strike of 1936–1937.* Ann Arbor: University of Michigan Press, 1969.

Flint Water Advisory Task Force. *Final Report.* March 2016. www.michigan.gov /documents/snyder/FWATF_FINAL_REPORT_21March2016_517805_7 .pdf.

Hayes, Chris. *A Colony in a Nation.* New York: W. W. Norton, 2017.

Highsmith, Andrew R. *Demolition Means Progress: Flint, Michigan, and the Fate of the American Metropolis.* Chicago: University of Chicago Press, 2015.

Jackson, Carlton. *Child of the Sit-Downs: The Revolutionary Life of Genora Dollinger.* Kent, Ohio: Kent State University Press, 2008.

Johnson, Steven. *The Ghost Map: The Story of London's Most Terrifying Epidemic—and How It Changed Science, Cities, and the Modern World.* New York: Riverhead Books, 2006.

Kitman, Jamie Lincoln. "The Secret History of Lead." *The Nation.* March 2, 2000.

Lichtenstein, Nelson. *Walter Reuther: The Most Dangerous Man in Detroit.* Champaign-Urbana: University of Illinois Press, 1997.

Markowitz, Gerald, and David Rosner. *Lead Wars: The Politics of Science and the Fate of America's Children.* Berkeley: University of California Press, 2013.

Michigan Civil Rights Commission. *The Flint Water Crisis: Systemic Racism Through the Lens of Flint.* February 2017. www.michigan.gov/documents/mdcr/VFlintCrisisRep-F-Edited3-13-17_554317_7.pdf.

Rothstein, Richard. *The Color of Law: A Forgotten History of How Our Government Segregated America.* New York: Liveright, 2017.

Sicherman, Barbara. *Alice Hamilton: A Life in Letters.* Champaign-Urbana: University of Illinois Press, 2003.

Skloot, Rebecca. *The Immortal Life of Henrietta Lacks.* New York: Crown, 2010.

Sugrue, Thomas J. *The Origins of the Urban Crisis: Race and Inequality in Postwar Detroit.* Princeton, N.J.: Princeton University Press, 1996.

Taylor, Dorceta E. *Toxic Communities: Environmental Racism, Industrial Pollution, and Residential Mobility.* New York: New York University Press, 2014.

Thomas, Hugh. *The Spanish Civil War.* New York: Modern Library, 2001.

Troesken, Werner. *The Great Lead Water Pipe Disaster.* Cambridge, Mass.: MIT Press, 2006.

Vinten-Johansen, Peter, Howard Brody, Nigel Paneth, Stephen Rachman, and Michael Rip, with David Zuck. *Cholera, Chloroform, and the Science of Medicine: A Life of John Snow.* New York: Oxford University Press, 2003.

Warren, Christian. *Brush with Death: A Social History of Lead Poisoning.* Baltimore: Johns Hopkins University Press, 2000.

THE FLINT CHILD HEALTH AND DEVELOPMENT FUND, FlintKids.org, was created by Dr. Mona and her partners as a "tomorrow fund" for Flint children. Housed at the Community Foundation of Greater Flint, the fund supports a myriad of interventions proven to promote children's potential: home visiting services, nutrition education, breastfeeding support, mindfulness programming, literacy efforts, play structures, and much more. A portion of the author's proceeds from this book will be donated to the fund. To consider donating, please visit FlintKids.org.

A LONG-TIME SUPPORTER of early literacy, Penguin Random House recognizes the role of reading in the successful development of all children, especially children struggling with adversities. Penguin Random House, through the ongoing operational partnership with the Imagination Library program, is supporting the home delivery of books for Flint kids 0 to 5 years of age.

Notes

═══

BELOW YOU'LL FIND A SELECTION OF SOURCES THAT INFORMED this book. Please see "Further Reading" and MonaHannaAttisha .com for more resources.

CHAPTER 1: WHAT THE EYES DON'T SEE

page 20: **"It's easier to build strong children than to repair broken men"**: This quote is frequently attributed to Douglass in sources that include the National Association for the Education of Young Children's newsletter, *Young Children* (volume 53), and *The New York Times* (www.nytimes.com/2014/03/01/opinion/blow-fathers-sons-and-brothers-keepers .html), but is not reliably sourced to Douglass's writings.

page 22: *The eyes don't see what the mind doesn't know*: The actual quote is "What the eye doesn't see and the mind doesn't know, doesn't exist." D. H. Lawrence, *Lady Chatterley's Lover* (first published in 1928, then suppressed until an epic thirty-year obscenity ban was overturned) (New York: Bantam Books, 1968), p. 16.

page 24: *"When we're sick, we hear"*: Bertolt Brecht, "A Worker's Speech to a Doctor" (1938), trans. Thomas Mark Kuhn and David J. Constantine, in *Collected Poems of Bertolt Brecht* (New York: Norton, 2018).

page 25: **chronically activates stress hormones**: Center on the Developing Child at Harvard University, *Toxic Stress*, 2017, developingchild.harvard.edu/science/key-concepts /toxic-stress.

page 25: **In a landmark study**: Vincent J. Felitti et al., "Relationship of Childhood Abuse and Household Dysfunction to Many of the Leading Causes of Death in Adults: The Adverse Childhood Experiences (ACE) Study," *American Journal of Preventive Medicine* 14, no. 4 (1998): 245–58.

page 25: **six or more ACEs**: David W. Brown et al., "Adverse Childhood Experiences and the Risk of Premature Mortality," *American Journal of Preventive Medicine* 37, no. 5 (2009): 389–96.

page 25:   **just one ACE puts a child:** Robyn Wing et al., "Association Between Adverse Childhood Experiences in the Home and Pediatric Asthma," *Annals of Allergy, Asthma & Immunology* 114, no. 5 (2015): 379–84.

page 26:   **the Tuskegee syphilis experiment:** Centers for Disease Control and Prevention, "U.S. Public Health Service Syphilis Study at Tuskegee: The Tuskegee Timeline," updated August 30, 2017, www.cdc.gov/tuskegee/timeline.htm.

CHAPTER 3: THE VALEDICTORIAN

page 41:   **But once it was in her bloodstream:** The scientific references regarding lead pathophysiology and impact are many; only a few are highlighted here. In 2006, the U.S. Environmental Protection Agency released a report on the nature and impacts of human lead exposure: U.S. EPA, *Air Quality Criteria for Lead (Final Report, 2006)*, EPA/600/R-05/144aF-bF (Washington, D.C.: Environmental Protection Agency, 2006). For comprehensive reviews of studies on the neurotoxic effects of lead on both children and adults, see Theodore I. Lidsky and Jay S. Schneider, "Lead Neurotoxicity in Children: Basic Mechanisms and Clinical Correlates," *Brain* 126, no. 1 (2003): 5–19, doi: 10.1093/brain/awg014; and Talia Sanders et al., "Neurotoxic Effects and Biomarkers of Lead Exposure: A Review," *Reviews on Environmental Health* 24, no. 1 (2009): 15–45. For an in-depth analysis of the effects found in adults and children with low-level lead exposure, see National Toxicology Program, *NTP Monograph on Health Effects of Low-Level Lead* (Washington, D.C.: Department of Health and Human Services, 2012), ntp.niehs.nih.gov/ntp/ohat/lead/final/monographhealtheffectslowlevellead_newissn_508.pdf. Even trace amounts of lead can impact children and adults; see National Institute of Environmental Health Services, "Environmental Agents: Lead," reviewed June 15, 2017, www.niehs.nih.gov/health/topics/agents/lead/index.cfm.

CHAPTER 5: RED FLAGS

page 55:   **In July 2015 he posted a news story:** Curt Guyette, "Scary: Leaded Water and One Flint Family's Toxic Nightmare," *Deadline Detroit,* July 9, 2015, www.deadlinedetroit.com/articles/12697/scary_leaded_water_and_one_flint_family_s_toxic_nightmare.

page 55:   **an eight-page interim report:** Miguel A. Del Toral, "High Lead Levels in Flint, Michigan—Interim Report" [memorandum], June 24, 2015, flintwaterstudy.org/wp-content/uploads/2015/11/Miguels-Memo.pdf.

page 59:   **"a rogue employee":** Sarah Hulett, "High Lead Levels in Michigan Kids after City Switches Water Source," NPR, September 29, 2015, www.npr.org/2015/09/29/444497051/high-lead-levels-in-michigan-kids-after-city-switches-water-source.

page 63:   **The D.C. water story broke:** Josh Levin, "Plumbing the Depths: The EPA Finds Too Much Lead in D.C. Tap Water," *Washington City Paper,* October 18, 2002, www.washingtoncitypaper.com/news/article/13025198/plumbing-the-depths.

page 64:   **"would literally have to be classified":** Pierre Home-Douglas, "The Water Guy," *Prism: American Society for Engineering Education* 14, no. 3 (2004), www.prism-magazine.org/nov04/feature_water.cfm.

page 64:   **In January 2004 *The Washington Post* published:** David Nakamura, "Water in D.C. Exceeds EPA Lead Limit," *Washington Post,* January 31, 2004, www.washingtonpost

.com/archive/politics/2004/01/31/water-in-dc-exceeds-epa-lead-limit/1e54ff9b-a393-4f0a
-a2dd-7e8ceeddre9r/.

page 65: **That wouldn't change until:** Reduction of Lead in Drinking Water Act, Pub. L.
No. 111-380, 124 STAT. 4131 (2011).

page 65: **cooked up its own corrupt study:** Centers for Disease Control and Prevention
(CDC), "Blood Lead Levels in Residents of Homes with Elevated Lead in Tap Water—
District of Columbia, 2004," *MMWR. Morbidity and Mortality Weekly Report* 53, no. 12
(2004): 268–70.

page 65: **in a partial replacement:** Miguel A. Del Toral, Andrea Porter, and Michael R.
Schock, "Detection and Evaluation of Elevated Lead Release from Service Lines: A Field
Study," *Environmental Science and Technology* 47, no. 16 (2013): 9300–9307.

page 65: **The following year he published a report:** Marc Edwards, Simoni Triantafyllidou, and Dana Best, "Elevated Blood Lead in Young Children Due to Lead-
Contaminated Drinking Water: Washington, DC, 2001–2004," *Environmental Science &
Technology* 43, no. 5 (2009): 1618–23, doi: 10.1021/es802789w. For details of the efforts required to convince authorities that children's blood lead levels had increased at least in part
due to the elevated concentrations of lead in their drinking water, see Robert McCartney,
"Virginia Tech Professor Uncovered Truth About Lead in D.C. Water," *Washington Post*,
May 23, 2010, www.washingtonpost.com/wp-dyn/content/article/2010/05/22/AR201005
2203447.html.

page 66: **"scientifically indefensible":** Carol D. Leonnig, "CDC Misled District Residents About Lead Levels in Water, House Probe Finds," *Washington Post*, May 20, 2010,
www.washingtonpost.com/wp-dyn/content/article/2010/05/19/AR2010051902599.html;
Darryl Fears, "GAO to Rebuke CDC for Playing Down Health Risk from Lead in D.C.
Tap Water," *Washington Post*, April 3, 2011, www.washingtonpost.com/national/gao-to
-rebuke-cdc-for-playing-down-health-risk-from-lead-in-dc-tap-water/2011/04/01
/AFvWkaXC_story.html.

CHAPTER 6: FIRST ENCOUNTER

page 71: **his angry, hate-spewing, anti-Semitic radio program:** "The Radio Priest,"
American Experience documentary, 1988; Sheldon Marcus, *Father Coughlin: The Tumultuous
Life of the Priest of the Little Flower* (Boston: Little, Brown, 1972).

page 72: **in 1934 Coughlin received:** Alan Brinkley, *Voices of Protest: Huey Long, Father
Coughlin & the Great Depression* (New York: Alfred A. Knopf, 1982), p. 119.

CHAPTER 7: MIASMA

page 83: **My favorite sleuth is John Snow:** There are many excellent John Snow refer-
ences, including these: Peter Vinten-Johansen et al., *Cholera, Chloroform, and the Science of
Medicine: A Life of John Snow* (New York: Oxford University Press, 2003), and its online
companion, johnsnow.matrix.msu.edu/index.php; Steven Johnson, *The Ghost Map: The
Story of London's Most Terrifying Epidemic—and How It Changed Science, Cities, and the
Modern World* (New York: Riverhead Books, 2006); Marjorie Bloy, "Cholera Comes to
Britain: October 1831," *A Web of English History,* modified March 4, 2016, www.history
home.co.uk/peel/p-health/cholera3.htm; and Ralph R. Frerichs, "John Snow: Removal of

the Pump Handle," UCLA Department of Epidemiology, www.ph.ucla.edu/epi/snow /removal.html.

page 87: **a bacteriologist named Paul Shekwana:** Sources about Paul Shekwana include these, some of which have been digitized by the Library of Congress: Dr. Shekwana appointment: *Iowa City Daily Press,* September 3, 1904, p. 1; offices in the third-floor medical building: *Iowa City Daily Press,* October 25, 1904, p. 8; work preventing typhoid: *Iowa City Daily Press,* May 16, 1905, p. 1; work with Iowa water supply: *Iowa City Daily Press,* June 26, 1906, p. 5; Dr. Shekwana's return to England: *Fredericksburg News,* Fredericksburg, Iowa, July 19, 1906, p. 3; his death: "Famous Scientist Is Killed at Iowa City," *Sioux Valley News,* Correctionville, Woodbury County, Iowa, July 12, 1906, p. 1; Dr. Shekwana's death in national news: "Crowded Over a Cliff," *Daily Nevada State Journal,* Reno, July 8, 1906, p. 1, and "Scientist Accidently Killed," *San Francisco Call,* July 8, 1906, p. 29; *JAMA* article: Paul Shekwana, "Disinfection of Physician's Hands," *Journal of the American Medical Association* 47, no. 2120 (1906).

page 90: **"In the little world in which children have their existence":** Charles Dickens, *Great Expectations* (New York: Penguin Classics, 1996), p. 63.

page 92: **"Flint drinking water meets":** Kristen Jordan Shamus, "State DEQ Didn't Take Flint Water Concerns Seriously," *Detroit Free Press,* February 14, 2016, www.freep .com/story/news/local/michigan/2016/02/13/state-deq-flint-water-concerns/80332954/.

pages 98–99: **"You don't want the higher chloride":** Ron Fonger, "General Motors Shutting Off Flint River Water at Engine Plant Over Corrosion Worries," *MLive,* October 13, 2014 (updated January 17, 2015), www.mlive.com/news/flint/index.ssf/2014/10/general_motors_wont_use_flint.html.

page 99: **"Let me start here":** Lindsey Smith, "Leaked Internal Memo Shows Federal Regulator's Concerns About Lead in Flint's Water," Michigan Radio, July 13, 2015, michi ganradio.org/post/leaked-internal-memo-shows-federal-regulator-s-concerns-about -lead-flint-s-water.

CHAPTER 8: NO RESPONSE

page 100: **a Nestorian stele:** Weam Namou, "The China Connection: Nestorian Stele Tells Ancient Tale," *Chaldean News,* December 26, 2015, culturalglimpse.com/2016/01/09 /the-china-connection-nestorian-stele-tells-ancient-tale/.

CHAPTER 9: SIT DOWN

page 118: **Long before cars were made:** My sources on Flint history include: Lawrence R. Gustin, *Billy Durant: Creator of General Motors* (Ann Arbor: University of Michigan Press, 1973); David L. Lewis, *The Public Image of Henry Ford: An American Folk Hero and His Company* (Detroit: Wayne State University Press, 1976); Andrew R. Highsmith, *Demolition Means Progress: Flint, Michigan, and the Fate of the American Metropolis* (Chicago: University of Chicago Press, 2015); R. C. Sadler and D. J. Lafreniere, "Racist Housing Practices as a Precursor to Uneven Neighborhood Change in a Post-Industrial City," *Housing Studies* 32, no. 2 (2017): 186–208; Thomas J. Sugrue, *The Origins of the Urban Crisis* (Princeton, N.J.: Princeton University Press, 1996); Kim Crawford, *The Daring Trader:*

Jacob Smith in the Michigan Territory, 1802–1825 (East Lansing: Michigan State University Press, 2012); Sidney Fine, *Sit-Down: The General Motors Strike of 1936–37* (Ann Arbor: University of Michigan Press, 1969); Nelson Lichtenstein, *Walter Reuther: The Most Dangerous Man in Detroit* (Champaign: University of Illinois Press, 1997); Walter P. Reuther Library, Wayne State University Archives (labor union oral history and image galleries); Flint Sit-Down website and links: reuther.wayne.edu/node/7092.

page 123: **Genora Johnson Dollinger, a serious socialist:** Susan Rosenthal, "Genora (Johnson) Dollinger Remembers the 1936–37 General Motors Sit-Down Strike," www.historyisaweapon.com/defcon1/dollflint.html; Studs Terkel, "Genora Johnson Dollinger," in *The Studs Terkel Reader: My American Century* (New York: New Press, 1997), pp. 511–20; *The Great Sit-Down: Yesterday's Witness in America,* BBC (1976), documentary; *With Babies and Banners: The Story of the Women's Emergency Brigade,* dir. Lorraine Gray (1979), Academy Award–nominated documentary.

page 125: **"The law knows no finer hour":** *Falbo v. United States,* 320 U.S. 549, 561 (1944), Justice Murphy dissenting opinion; quoted in J. Woodford Howard, Jr., *Mr. Justice Murphy: A Political Biography* (Princeton, N.J.: Princeton University Press, 1968), p. 33.

page 125: **"legalization of racism":** *Korematsu v. United States,* 323 U.S. 214, 242 (1944), Justice Murphy dissenting opinion; quoted in Robert Havey, "The Dissenter," bentley.umich.edu/news-events/magazine/the-dissenter/.

page 125: **Dr. Ossian Sweet:** Kevin Boyle, *Arc of Justice: A Saga of Race, Civil Rights, and Murder in the Jazz Age* (New York: Henry Holt, 2004).

page 126: **an important test of justice:** Joseph Turrini, "Sweet Justice," *Michigan History Magazine,* July–August 1999, pp. 22–27, www.michigan.gov/documents/dnr/mhc_mag_sweet-justice_308404_7.pdf.

page 126: **outlawed racial housing covenants:** *Shelly v. Kraemer,* 334 U.S. 1 (1948).

page 126: **1954** *Brown v. Board of Education* **decision:** *Brown v. Board of Education,* 347 U.S. 483 (1954).

page 126: **riots in Detroit and Flint:** "See How Riots in Detroit 50 Years Ago Spread to Flint," *MLive,* July 24, 2017, www.mlive.com/news/flint/index.ssf/2017/07/see_how_riots_in_detroit_50_ye.html.

page 126: **a groundbreaking fair housing ordinance:** Joe Lawlor, "Flint Made Civil Rights History 40 Years Ago," *Flint Journal,* February 10, 2008, blog.mlive.com/flintjournal/newsnow/2008/02/flint_made_civil_rights_histor.html.

page 127: **"Court's refusal to remedy":** *Milliken v. Bradley,* 418 U.S. 717 (1974), Justice Marshall dissenting opinion.

page 128: **almost 60 percent of children live in poverty:** The 2017 Flint child poverty rate was 58.3 percent. Michigan League for Public Policy, "Flint: 2017 Trends in Child Well-Being," *2017 Kids Count in Michigan Data Book,* www.mlpp.org/wp-content/uploads/2017/04/Flint-db-2017-Rev.pdf.

page 128: **48 killings in Flint:** Jake May, "Six Startling Statistics about Guns and Homicides in Flint," *MLive,* March 10, 2016, www.mlive.com/news/flint/index.ssf/2016/03/six_startling_statistics_of_gu.html.

page 128: **state of Michigan cut revenue sharing:** Dominic Adams, "Report Says Flint Lost Out on Nearly $55 Million in Revenue Sharing in Last Decade," *MLive,* March 19, 2014, www.mlive.com/news/flint/index.ssf/2014/03/report_says_flint_lost_out_on.html;

Mitch Bean, "Mitch Bean: Starving Michigan Cities and the Coming Storm," *MLive,* June 1, 2016, www.mlive.com/opinion/index.ssf/2016/06/mitch_bean_starving_michigan _c.html.

page 129: **Public Act 4:** Josh Hakala, "How Did We Get Here? A Look Back at Michigan's Emergency Manager Law," Michigan Radio, February 3, 2016, michiganradio.org /post/how-did-we-get-here-look-back-michigans-emergency-manager-law.

page 130: **Between 2008 and 2016, it shrank:** Alex Kellogg, "What Flint's Water Crisis and Its Gun Violence Epidemic Have in Common," *Trace,* March 16, 2016, www.thetrace .org/2016/03/what-flints-water-crisis-and-its-gun-violence-epidemic-have-in-common/.

CHAPTER 10: JENNY + THE DATA

page 134: **"Marc Edwards, a civil engineer":** MacArthur Foundation, "Marc Edwards: Water Quality Engineer | Class of 2007," updated August 2015, www.macfound.org /fellows/823/.

page 143: **All research done on humans is protected:** The Tuskegee Study resulted in the passage of the National Research Act of 1974 and the establishment of a Health and Human Services Policy for Protection of Human Research Subjects. As a result, all U.S. research involving human subjects must now be reviewed and approved by an Institutional Review Board (IRB). Wayne W. LaMorte, "Institutional Review Boards and the Belmont Principles: The Syphilis Study at Tuskegee," modified June 8, 2016, sphweb.bumc.bu.edu /otlt/mph-modules/ep/ep713_researchethics/ep713_researchethics3.html; Centers for Disease Control and Prevention (CDC), "U.S. Public Health Service Syphilis Study at Tuskegee: Research Implications," updated February 22, 2017, www.cdc.gov/tuskegee/after.htm.

page 144: **"They pull that rabbit out":** Ron Fonger, "Feds Sending in Experts to Help Flint Keep Lead Out of Water," *MLive,* September 10, 2015, www.mlive.com/news/flint /index.ssf/2015/09/university_researchers_dont_dr.html.

CHAPTER 11: PUBLIC HEALTH ENEMY #1

page 146: **Lead is probably the most widely studied neurotoxin:** There are many terrific books and magazine articles on the history of industrial lead use. I learned most from: Lydia Denworth, *Toxic Truth: A Scientist, a Doctor, and the Battle over Lead* (Boston: Beacon Press, 2009); Gerald Markowitz and David Rosner, *Lead Wars: The Politics of Science and the Fate of America's Children* (Berkeley: University of California Press, 2013); Christian Warren, *Brush with Death: A Social History of Lead Poisoning* (Baltimore: Johns Hopkins University Press, 2000); Jamie Lincoln Kitman, "The Secret History of Lead," *Nation,* March 2, 2000, www.thenation.com/article/secret-history-lead/; William Kovarik, "Ethyl-Leaded Gasoline: How a Classic Occupational Disease Became an International Public Health Disaster," *International Journal of Occupational and Environmental Health* 11, no. 4 (2005): 384–97.

page 146: **lead in the Romans' food:** The debate about the role of lead in the fall of Rome has been raging for decades. It started with Jerome O. Nriagu, "Saturnine Gout Among Roman Aristocrats: Did Lead Poisoning Contribute to the Fall of the Empire?," *New England Journal of Medicine* 308, no. 11 (1983): 660–63.

page 147: **"gift from God":** "A 'Gift from God'? The Public Health Controversy over

Leaded Gasoline During the 1920s," *American Journal of Public Health* 75, no. 4 (1985): 344–52; and *Deceit and Denial: The Deadly Politics of Industrial Pollution* (Berkeley, Calif.: Milbank Memorial Fund, 2002), pp. 12–35.

page 149: **Alice Hamilton, who lived:** Biographical sources include: Barbara Sicherman, *Alice Hamilton: A Life in Letters* (Champaign-Urbana: University of Illinois Press, 2003); on Hamilton's time as a medical student at University of Michigan and later joining the Harvard faculty: American Chemical Society National Historic Chemical Landmarks, "Alice Hamilton and the Development of Occupational Medicine," updated November 5, 2015, www.acs.org/content/acs/en/education/whatischemistry/landmarks/alicehamilton .html; as a professor of pathology at the Woman's Medical School of Northwestern University: U.S. National Library of Medicine, "Biography: Dr. Alice Hamilton," *Changing the Face of Medicine,* updated June 3, 2015, cfmedicine.nlm.nih.gov/physicians/biography _137.html; at Hull House: Jane Addams, *Twenty Years at Hull-House: with Autobiographical Notes* (New York: Macmillan, 1912); her epic battle against Kettering and Ethyl Corporation: William Kovarik, "Charles F. Kettering and the 1921 Discovery of Tetraethyl Lead," paper to the Society of Automotive Engineers, Fuels and Lubricants Division conference, Baltimore, 1994, www.environmentalhistory.org/billkovarik/about-bk/research/cabi /ket-tel/#early, and Kovarik, "The Ethyl Conflict & the Media," paper to the Association for Education in Journalism and Mass Communication, April 1994, www.environmental history.org/billkovarik/about-bk/research/cabi/the-ethyl-conflict/.

page 151: **He insisted that lead was naturally occurring:** Herbert L. Needleman, "The Removal of Lead from Gasoline: Historical and Personal Reflections," *Environmental Research* 84, no. 1 (2000): 20–35.

page 153: **The lead industry used its towering advantage:** Gerald Markowitz and David Rosner, " 'Cater to the Children': The Role of the Lead Industry in a Public Health Tragedy, 1900–1955," *American Journal of Public Health* 90, no. 1 (2000): 36–46.

page 153: **12.4 percent of the global burden:** World Health Organization, "Lead Poisoning and Health: Fact Sheet," updated August 2017, www.who.int/mediacentre/factsheets /fs379/en/. See also Bruce Lanphear et al., "Low-level Lead Exposure and Mortality in U.S. Adults: A Population-Based Cohort Study," *Lancet Public Health* (2018): doi: 10.1016 /S2468-2667(18)30025-2.

page 155: **three Arab countries—Algeria, Yemen, and Iraq:** Partnership for Clean Fuels and Vehicles, "Meeting Report: 11th PCFV Global Partnership Meeting, 6–7 June 2016, London," staging.unep.org/Transport/new/PCFV/pdf/11GPM_WorkshopReport .pdf; United Environment Programme, *Leaded Petrol Phase-out: Global Status as at March 2017* (Nairobi: UNEP, 2017), wedocs.unep.org/bitstream/handle/20.500.11822/17542/Map WorldLead_March2017.pdf?sequence=1&isAllowed=y.

page 155: **we know lead's potential:** David Bellinger, "Very Low Lead Exposures and Children's Neurodevelopment," *Current Opinion in Pediatrics* 20 no. 2 (2008): 172–77, doi: 10.1097/MOP.0b013e3282f4f97b.

page 155: **Econometrics studies:** Several studies have correlated lead exposure with crime: Rick Nevin, "How Lead Exposure Relates to Temporal Changes in IQ, Violent Crime, and Unwed Pregnancy," *Environmental Research* 83, no. 1 (2000): 1–22. Analyses indicate a significant relationship between childhood lead exposure and violent crime in adulthood. See Jessica Wolpaw Reyes, "Environmental Policy as Social Policy? The Impact of Childhood Lead Exposure on Crime," *B.E. Journal of Economic Analysis and Policy*

(2007): article 51; and John Paul Wright et al., "Association of Prenatal and Childhood Blood Lead Concentrations with Criminal Arrests in Early Adulthood," *PLOS Medicine* 5, no. 5 (2008): e101. Studying preschool blood-lead levels and national crime rate trends, Nevin found a strong association between the two in nine countries. See Rick Nevin, "Understanding International Crime Trends: The Legacy of Preschool Lead Exposure," *Environmental Research* 104, no. 3 (2007): 315–36.

page 156: **But here's something beautiful:** American Academy of Pediatrics Council on Environmental Health, "Prevention of Childhood Lead Toxicity," *Pediatrics* 138, no. 1 (2016): e20161493, doi: 10.1542/peds.2016-1493.

page 158: **I learned that in the last 150 years:** Werner Troesken, *The Great Lead Water Pipe Disaster* (Cambridge, Mass.: MIT Press, 2006).

page 158: **a lead-based abortion pill:** Alison Moulds, "The Other Side to the Story: Abortion and Family Planning," Victorian Clinic, June 29, 2013, victorianclinic.wordpress .com/2013/06/29/the-other-side-to-the-story-abortion-and-family-planning/.

CHAPTER 13: THE MAN IN THE PANDA TIE

page 178: **Lansing had elevated lead levels in its water:** Ed Glaser, "Weighing Last Summer's Lansing Lead Scare," *City Pulse,* January 5, 2005, lansingcitypulse.com/archives /050105/features/index2.asp.

CHAPTER 15: POISONED BY POLICY

page 204: **Citizens went to the UN:** Curt Guyette, "In Flint, Michigan, Overpriced Water Is Causing People's Skin to Erupt in Rashes and Hair to Fall Out," *Nation,* July 6, 2015, www.thenation.com/article/in-flint-michigan-overpriced-water-is-causing-peoples -skin-to-erupt-and-hair-to-fall-out/.

CHAPTER 16: SHORTWAVE RADIO CRACKLING

page 210: **But now, because of my dad's discovery:** Weam Namou, "It Takes Two Villages," *Chaldean News,* January 2012, p. 40, issuu.com/chaldeannews/docs/20120104155420819/9.

page 211: **Israel Raba and his family were famous scribes:** David Wilmshurst, *Ecclesiastical Organization of the Church of the East, 1318–1913* (n.p.: Corpus Scriptorum Christanorum Orientalium, 2000). The history of the Assyrians and of the interactions between the Church of the East and the Chaldean church is fraught and frightfully complex. My few sentences do not do them justice. The centuries of plagues, massacres, and invasions only underscore my respect for the religious and cultural resilience of all the great people of the region.

page 212: **invaluable artifacts and religious books:** Muna Fadhil, "Isis Destroys Thousands of Books and Manuscripts in Mosul Libraries," *Guardian,* February 26, 2015, www .theguardian.com/books/2015/feb/26/isis-destroys-thousands-books-libraries.

page 212: **in the ferment of the Great Depression:** Hanna Batatu, *The Old Social Classes and the Revolutionary Movements of Iraq: A Study of Iraq's Old Landed and Commercial Classes and of its Communists, Ba'thists, and Free Officers* (Princeton, N.J.: Princeton University Press, 1983).

page 213: joining 35,000 other freedom-loving idealists: Hugh Thomas, *The Spanish Civil War* (New York: Modern Library, 2001); and Salvador Bofarull, "Brigadistas árabes en la Guerra de España: Combatientes por la República," *Nacion Arabe* 52 (2004): 121–34.

page 213: General Motors and other U.S. auto companies: Adam Hochschild, *Spain in our Hearts: Americans in the Spanish Civil War, 1936–1939* (New York: Houghton Mifflin Harcourt, 2016).

page 217: the United States actively supported Saddam: Joyce Battle, ed., "Shaking Hands with Saddam Hussein: The U.S. Tilts Toward Iraq, 1980–1984," *National Security Archive Electronic Briefing Book No. 82* (Washington, D.C.: National Security Archive, 2003), nsarchive2.gwu.edu//NSAEBB/NSAEBB82/index.htm; Alan Friedman, *Spider's Web: The Secret History of How the White House Illegally Armed Iraq,* (New York: Bantam, 1993).

CHAPTER 17: MEETING THE MAYOR

page 225: "Because no measurable level": Centers for Disease Control and Prevention (CDC), Advisory Committee on Childhood Lead Poisoning Prevention, *Low Level Lead Exposure Harms Children: A Renewed Call for Primary Prevention* (Atlanta: CDC, January 4, 2012), www.cdc.gov/nceh/lead/acclpp/final_document_030712.pdf.

page 227: "For childhood lead poisoning": L. Trasande and Y. Liu, "Reducing the Staggering Costs of Environmental Disease in Children, Estimated at $76.6 Billion in 2008," *Health Affairs* 30, no. 5 (2011): 863–70, doi: 10.1377/hlthaff.2010.1239.

page 227: it showed economic losses attributable: Michigan Network for Children's Environmental Health, and the Ecology Center, "The Price of Pollution: Cost Estimates of Environment-Related Childhood Disease in Michigan," June 2010, www.sehn.org/tccpdf/childnood%20illness.pdf.

page 227: For about 25% of infants: Simoni Triantafyllidou, Daniel Gallagher, and Marc Edwards, "Assessing Risk with Increasingly Stringent Public Health Goals: The Case of Water Lead and Blood Lead in Children," *Journal of Water and Health* 12, no. 1 (2014): 57–68.

page 227: Increase in fetal death: Marc Edwards, "Fetal Death and Reduced Birth Rates Associated with Exposure to Lead-Contaminated Drinking Water," *Environmental Science and Technology* 48, no. 1 (2014): 739–46.

CHAPTER 19: THE PRESS CONFERENCE

page 250: Ron Fonger wrote a story: Ron Fonger, "Doctors to Speak Out Today on Lead in Flint Water," *MLive,* September 24, 2015, www.mlive.com/news/flint/index.ssf/2015/09/doctors_health_officials_to_sp.html.

page 253: "It transforms fidelity into infidelity": Karl Marx, *Economic and Philosophic Manuscripts of 1844: The Power of Money,* www.marxists.org/archive/marx/works/1844/manuscripts/power.htm.

CHAPTER 20: SPLICE AND DICE

page 259: The blowback began immediately: Ron Fonger, "State Says Data Shows No Link to Flint River, Elevated Lead in Blood," *Flint Journal,* September 24, 2015, www

.mlive.com/news/flint/index.ssf/2015/09/state_says_its_data_shows_no_c.html. The article opens: "The state is wasting no time in taking issue with a study linking use of the Flint River as a drinking water source to elevated blood lead levels in children."

page 259: "fanning political flames irresponsibly": Marc Edwards and Siddhartha Roy, "Commentary: MDEQ Mistakes and Deception Created the Flint Water Crisis," September 30, 2015, flintwaterstudy.org/2015/09/commentary-mdeq-mistakes-deception-flint -water-crisis/; September 8, 2015, email from Brad Wurfel to Ron Fonger, senatedems .com/snyder_emails/20150901_September%201%20-%2024%2C%202015.pdf.

page 260: "I would call them *unfortunate*": Associated Press, "Did This Michigan Town Poison Its Children?: Children in Flint, Michigan Are Showing High Levels of Blood Lead," *U.S. News & World Report,* September 25, 2015, www.usnews.com/news/articles /2015/09/25/flint-michigan-children-show-high-levels-of-lead-in-blood.

page 260: "seasonal anomaly": Editorial Board, "Lead-Poisoned Flint Kids Will Pay Price for Water Switch," *Detroit Free Press,* September 24, 2015, www.freep.com/story /opinion/editorials/2015/09/24/lead-poisoned-flint-kids-pay-price-water-switch/7274 1222/; Robin Erb, "Doctor: Lead Seen in More Flint Kids Since Water Switch," *Detroit Free Press,* September 24, 2015, www.freep.com/story/news/local/michigan/2015/09/24 /water-lead-in-flint/72747696/.

page 261: "spliced and diced": Nancy Kaffer, "Snyder Must Act on Flint Lead Crisis," *Detroit Free Press,* September 27, 2015, www.freep.com/story/opinion/columnists/nancy -kaffer/2015/09/26/shortfalls-em-law-writ-large-flint-water-crisis/72811990/; email from Sara Wurfel to Nancy Kaffer, September 24, 2015, saying data is "spliced and diced," senatedems.com/snyder_emails/20150901_September%201%20-%2024%2C%202015.pdf, pp. 828–29.

CHAPTER 21: NUMBERS WAR

page 269: "Data that the state of Michigan": Kristi Tanner and Nancy Kaffer, "State Data Confirms Higher Blood-Lead Levels in Flint Kids," *Detroit Free Press,* September 29, 2015, www.freep.com/story/opinion/columnists/nancy-kaffer/2015/09/26/state-data -flint-lead/72820798/.

page 270: "It's hard to understand the resounding yawn": Nancy Kaffer, "Snyder Must Act on Flint Lead Crisis," *Detroit Free Press,* September 27, 2015, www.freep.com/story /opinion/columnists/nancy-kaffer/2015/09/26/shortfalls-em-law-writ-large-flint-water -crisis/72811990/.

page 271: an "ecobiodevelopmental" approach: Jack P. Shonkoff et al., "The Lifelong Effects of Early Childhood Adversity and Toxic Stress," *Pediatrics* 129, no. 1 (2011): e232–e246, doi: 10.1542/peds.2011-2663.

page 272: our friends in the north: "Flint, Michigan, Declares Emergency; High Lead Levels in Kids Linked to Tap Water," *As It Happens with Carol Off and Jeff Douglas,* Canadian Broadcasting Company, October 2, 2015, www.cbc.ca/radio/asithappens/as-it-hap pens-friday-edition-1.3254263/flint-michigan-declares-emergency-after-doctor-links -children-to-lead-tainted-tap-water-1.3254267.

page 272: *All Things Considered:* Sarah Hulett, "High Lead Levels in Michigan Kids After City Switches Water Source," *All Things Considered,* NPR, September 29, 2015, www

.npr.org/2015/09/29/444497051/high-lead-levels-in-michigan-kids-after-city-switches
-water-source.

CHAPTER 22: DEMONSTRATION OF PROOF

page 277:   **I shared recent research from Montreal:** Gerard Ngueta et al., "Use of a Cumulative Exposure Index to Estimate the Impact of Tap Water Lead Concentration on Blood Lead Levels in 1- to 5-Year-Old Children (Montréal, Canada)," *Environmental Health Perspectives* 124, no. 3 (2016): 388–95, doi: 10.1289/ehp.1409144.

page 281:   **"We understand many have lost confidence":** Robin Erb, "Flint Doctor Makes State See Light about Lead in Water," *Detroit Free Press,* October 12, 2015, www.freep.com/story/news/local/michigan/2015/10/10/hanna-attisha-profile/73600120/.

page 281:   **ten-point "comprehensive action" plan:** Office of Governor Rick Snyder, "Comprehensive Action Plan Will Help Flint Residents Address Water Concerns" [press release], October 2, 2015, www.michigan.gov/snyder/0,4668,7-277-57577_57657-366315--,00.html.

CHAPTER 23: ALL THE THINGS WE FOUND OUT LATER

page 284:   **arranging for water coolers to be delivered:** Paul Egan, "Amid Denials, State Workers in Flint Got Clean Water," *Detroit Free Press,* January 28, 2016, www.freep.com/story/news/local/michigan/flint-water-crisis/2016/01/28/amid-denials-state-workers-flint-got-clean-water/79470650/.

page 284:   **exploring the distribution of water filters:** Jonathan Oosting and Chad Livengood, "Snyder Aides Considered Flint Water Filters in March," *Detroit News,* February 12, 2016, www.detroitnews.com/story/news/michigan/flint-water-crisis/2016/02/12/snyder-aides-considered-flint-water-filters-march/80270588/.

page 285:   **it would have cost only eighty dollars a day:** Claire Bernish, "Flint Officials Could Have Prevented Lead Crisis for $80 a Day," *Mint Press News,* February 5, 2016, www.mintpressnews.com/flint-officials-could-have-prevented-lead-crisis-for-80-a-day/213462/.

page 285:   **the MDHHS had an analysis done:** Vanessa Schipani, "False Claims About Flint Water," FactCheck, April 27, 2016, www.factcheck.org/2016/04/false-claims-about-flint-water/.

page 286:   **"Yes, the issue is moving":** *Bridge* magazine staff, *Poison on Tap: How Government Failed Flint, and the Heroes Who Fought Back* (Traverse City, Mich.: Mission Point Press, 2016), p. 110.

page 286:   **strange escalation in cases of Legionnaires' disease:** Scientific and media articles related to the Flint water crisis and Legionnaires' disease and pneumonia rates: William J. Rhoads et al., "Distribution System Operational Deficiencies Coincide with Reported Legionnaires' Disease Clusters in Flint, Michigan," *Environmental Science and Technology* 51, no. 20 (2017): 11986–95; David Otto Schwake et al., "*Legionella* DNA Markers in Tap Water Coincident with a Spike in Legionnaires' Disease in Flint, MI," *Environmental Science and Technology* 3, no. 9 (2016): 311–15; Ron Fonger, "CDC Finds First Genetic Link Between Legionnaires' Outbreak, Flint Water," *MLive,* February 16, 2017, www.mlive.com/news/flint/index.ssf/2017/02/cdc_finds_first_genetic_link_b.html; in regard to an increase in pneumonia deaths, Chastity Pratt Dawsey, "Soaring Pneumonia Deaths in Genesee County Likely Linked to Undiagnosed Legionnaires', Experts Say," *Bridge* magazine, January 26, 2017, www.bridgemi.com/children-families/soaring -pneumonia-deaths-genesee-county-likely-linked-undiagnosed-legionnaires.

page 286: "I'm not so sure Flint": Stephanie Akin, "Was EPA Unwilling to 'Go Out on a Limb' for Flint? Committee Explores Federal, State Role in Water Crisis," *Roll Call,* March 16, 2016, www.rollcall.com/news/policy/was-epa-unwilling-to-go-out-on-a-limb -for-flint.

page 287: "obscene failure of government": Editorial Board, "Flint Water Crisis: An Obscene Failure of Government," *Detroit Free Press,* October 8, 2015, www.freep.com /story/opinion/editorials/2015/10/08/flint-water-crisis-obscene-failure-government/7357 8640/.

page 287: He even offered the assistance of the foundation: Jennifer Chambers, "Mott Foundation Hoping for 'Dramatic Results' in Flint," *Detroit News,* October 8, 2015, www .detroitnews.com/story/news/local/michigan/2015/10/08/mott-foundation-grant-flint -water/73599612/.

page 291–92: It eventually issued a directive to prevent: U.S. Environmental Protection Agency, "Flint Drinking Water Technical Support Team," updated January 23, 2018, www .epa.gov/flint/flint-drinking-water-technical-support-team.

CHAPTER 24: FIRE ANT

page 295: accepted for publication: M. Hanna-Attisha et al., "Elevated Blood Lead Levels in Children Associated with the Flint Drinking Water Crisis: A Spatial Analysis of Risk and Public Health Response," *American Journal of Public Health* 106, no. 2 (2016): 283–90, doi: 10.2105/AJPH.2015.303003.

page 299: The number forty-three was picked up again: Mitch Smith, "Flint Wants Safe Water, and Someone to Answer for Its Crisis," *New York Times,* January 9, 2016, www .nytimes.com/2016/01/10/us/flint-wants-safe-water-and-someone-to-answer-for-its -crisis.html.

page 299: the Kettering University president: Dave Bartkowiak, Jr., "Kettering University President Aims to Clarify, Detail Scope of Flint Water Crisis: University President Points to Disparity Among Pipes," *Click on Detroit,* January 21, 2016, www.clickondetroit .com/news/michigan/scope-of-flint-water-contamination-downplayed-calrified-in -kettering-university-presidents-letter.

page 299: she said I was "badass": Rachel Maddow, "Michigan's Snyder Pressed for Action on Flint," MSNBC, January 14, 2016, www.msnbc.com/transcripts/rachel-maddow -show/2016-01-13.

page 300: "consider the whole cohort": Kristi Tanner, "All Flint's Children Must Be Treated as Exposed to Lead," *Detroit Free Press,* January 16, 2016, www.freep.com/story /opinion/contributors/raw-data/2016/01/16/map-8657-flints-youngest-children-exposed -lead/78818888/.

page 301: PBB contamination in 1973: Joyce Egginton, *The Poisoning of Michigan,* 2nd revised edition (East Lansing: Michigan State University Press, 2009).

CHAPTER 25: TRUTH AND RECONCILIATION

page 307: **review the constitutionality of the EM law:** Todd Spangler, "U.S. Supreme Court Rejects Challenge to Michigan's Emergency Manager Law," *Detroit Free Press,* October 2, 2017, www.freep.com/story/news/local/michigan/2017/10/02/u-s-supreme -court-michigan-emergency-manager-law/723074001/; Chris Savage, "The Scandal of Michigan's Emergency Managers," *Nation,* February 15, 2012, www.thenation.com/article /scandal-michigans-emergency-managers/.

page 307: **"This is a racial crime":** Michael Moore, "Flint Poisoning Is a 'Racial Crime,'" *Time,* January 21, 2016, time.com/4188323/michael-moore-flint-racial-crime/.

page 308: **a kickass 112-page report:** Flint Water Advisory Task Force, "Final Report," March 2016, www.michigan.gov/documents/snyder/FWATF_FINAL_REPORT _21March2016_517805_7.pdf.

page 308: **the Michigan Department of Civil Rights made similar findings:** Michigan Civil Rights Commission, "The Flint Water Crisis: Systemic Racism Through the Lens of Flint," Report of the Michigan Civil Rights Commission, February 17, 2017, www.michi gan.gov/documents/mdcr/VFlintCrisisRep-F-Edited3-13-17_554317_7.pdf.

page 308: **"I guarantee you":** Daniel Bethencourt and Kathleen Gray, "Jesse Jackson Calls Flint Crisis 'a Crime Scene,'" *Detroit Free Press,* January 17, 2016, www.freep.com /story/news/local/michigan/flint-water-crisis/2016/01/17/jesse-jackson-flint-calls-crisis -crime-scene/78939208/.

page 311: **the first criminal charges were announced:** On charges related to the Flint water crisis, sources include: Michigan Department of Attorney General, "Schuette Charges MDHHS Director Lyon, Four Others with Involuntary Manslaughter in Flint Water Crisis," June 14, 2017, www.michigan.gov/ag/0,4534,7-359-82916_81983_47203 -390055--,00.html; Paul Egan, "These Are the 15 People Criminally Charged in the Flint Water Crisis," *Detroit Free Press,* June 14, 2017, www.freep.com/story/news/local/michigan /flint-water-crisis/2017/06/14/flint-water-crisis-charges/397425001/; Ron Fonger, "Man-slaughter Charge Reaches Gov. Snyder's Cabinet over Flint Water Crisis," *MLive,* June 14, 2017, www.mlive.com/news/flint/index.ssf/2017/06/two.html; Dominique Debucquoy-Dodley, "Did Michigan Officials Hide the Truth About Lead in Flint?" CNN, January 14, 2016, www.cnn.com/2016/01/14/us/flint-water-investigation/; Talia Buford, "In Flint Water Crisis, Could Involuntary Manslaughter Charges Actually Lead to Prison Time?" *ProPublica,* June 19, 2017, www.propublica.org/article/flint-water-crisis-involuntary-man slaughter-charges-lead-to-prison-time.

page 312: **"a strong statement with a demonstration":** Flint Water Crisis investigative team, Michigan Department of Attorney General, *Interim Report of the Flint Water Crisis Investigation,* p. 8, www.michigan.gov/documents/ag/Flint+Water+Interim+Report _575711_7.pdf.

page 312: **A scathing report by the agency's inspector general:** U.S. Environmental Protection Agency, Office of Inspector General, *Report: Drinking Water Contamination in Flint, Michigan, Demonstrates a Need to Clarify EPA Authority to Issue Emergency Orders to Protect the Public,* Report no. 17-P-0004, October 20, 2016, www.epa.gov/office -inspector-general/report-drinking-water-contamination-flint-michigan-demon strates-need.

page 312:   This prompted Representative Kildee: Water Infrastructure Improvements for the Nation Act, Pub. L. No. 114-322, 130 STAT. 1722–1726 (2016), www.gpo.gov/fdsys /pkg/BILLS-114s612enr/pdf/BILLS-114s612enr.pdf.

page 314:   Senator Stabenow, who had kept: Ron Fonger, "Lead-Exposed Flint Kids Need Better Nutrition Now, Stabenow Says," *MLive,* November 17, 2015, www.mlive.com /news/flint/index.ssf/2015/11/stabenow_says_shes_working_to.html.

Index

Image Credits

———

Title page: The border on the title page is adapted from the scribe priest Giwargis manuscript cover page, circa 1700. Courtesy of Chaldean Catholic Eparchy of Saint Thomas the Apostle of Detroit.

page 12: The fleuron ornament is a typographical illustration first printed in 1783's *Poetical Sketches* by William Blake.

page 16: Doug Pike, Hurley Medical Center

page 30: Samuel Wilson, used with permission of *The Flint Journal*

page 86: Snow, John. *On the Mode of Communication of Cholera*, 2nd ed. London: Churchill, 1855.

page 88: Courtesy of Hekmat Shikwana

page 121: Library of Congress Prints & Photographs Division, Washington, D.C. 20540, USA hdl.loc.gov/loc.pnp/pp.print

page 124: Used with permission of Walter P. Reuther Library, Archives of Labor and Urban Affairs, Wayne State University

page 135: Courtesy of John D. & Catherine T. MacArthur Foundation, www.macfound .org

page 150: Alice Hamilton student portrait, ca. 1893 (HS10419). J. J. Gibson, University of Michigan student portraits, Bentley Historical Library, University of Michigan

page 154: Reproduced and modified with permission from *Pediatrics*, vol. 38, p. 2, copyright © 2016 by the AAP

page 174: Stephen Carmody, Michigan Radio

page 211 (left and right): Courtesy of Iraq Centre Numérique des Manuscrits Orientaux (CNMO)

page 214: Courtesy of Salvador Bofarull/Amigos Brigadas Internacionales, from the files of the Russian State Archive of Socio-Political History, International Brigades Archive

page 219: Ahmad Nateghi

page 255: Doug Pike, Hurley Medical Center

page 257: Jake May, used with permission of *The Flint Journal*

page 297: Jake May, used with permission of *The Flint Journal*

page 311: Courtesy of Virginia Tech/Logan Wallace

page 319: Norman Bryant

ABOUT THE AUTHOR

MONA HANNA-ATTISHA is a physician, scientist, and public health activist. She has been awarded the Freedom of Expression Courage Award from PEN America, named to the *Time* 100, and called to testify before the United States Congress.

She's also the founder and director of the Pediatric Public Health Initiative, a model program to mitigate the impact of the Flint water crisis so that all Flint children grow up healthy and strong. Find out more about the organization and its work at MSUHurleyPPHI.org.

This is her first book.

MonaHannaAttisha.com

Twitter: @MonaHannaA

ABOUT THE TYPE

This book was set in Caslon, a typeface first designed in 1722 by William Caslon (1692–1766). Its widespread use by most English printers in the early eighteenth century soon supplanted the Dutch typefaces that had formerly prevailed. The roman is considered a "workhorse" typeface due to its pleasant, open appearance, while the italic is exceedingly decorative. The Declaration of Independence was first set into lead type and printed in Caslon.